T0275471

JAPAN'S NUCLEAR DISASTER AND THE POLITICS OF SAFETY GOVERNANCE

JAPAN'S NUCLEAR DISASTER AND THE POLITICS OF SAFETY GOVERNANCE

Florentine Koppenborg

CORNELL UNIVERSITY PRESS   ITHACA AND LONDON

First published 2023 by Cornell University Press

Library of Congress Cataloging-in-Publication Data

Names: Koppenborg, Florentine, 1986– author.
Title: Japan's nuclear disaster and the politics of safety governance / Florentine Koppenborg.
Description: Ithaca [New York] : Cornell University Press, 2023. | Includes bibliographical references and index.
Identifiers: LCCN 2022059016 (print) | LCCN 2022059017 (ebook) | ISBN 9781501770043 (hardcover) | ISBN 9781501770050 (pdf) | ISBN 9781501770067 (epub)
Subjects: LCSH: Genshiryoku Kisei Iinkai—Influence. | Fukushima Nuclear Disaster, Japan, 2011—Political aspects. | Nuclear power plants—Safety measures—Government policy—Japan. | Nuclear energy—Government policy—Japan. | Nuclear energy—Safety regulations—Japan.
Classification: LCC TK1365.J3 .K665 2023 (print) | LCC TK1365.J3 (ebook) | DDC 363.17/990952117—dc23/eng/20230103
LC record available at https://lccn.loc.gov/2022059016
LC ebook record available at https://lccn.loc.gov/2022059017

To my mother and my grandfather

Contents

Acknowledgments

This book is the result of a long process during which I have received much support and help. Going back all the way to my PhD thesis, I am grateful to my three supervisors—Miranda Schreurs, Verena Blechinger-Talcott, and Greg Noble—for their support and for encouraging me to look at my research from different viewpoints. The Graduate School of East Asian Studies (GEAS) at the Freie Universität Berlin provided the best PhD research environment I could ask for, also due to the graduate school coordinator Katrin Gengenbach who went out of her way to provide help whenever needed. I am indebted to all my fellow PhD students for their support, feedback, and friendship, but especially Ulv Hanssen for tirelessly engaging in discussions and for unwinding with good food, drinks, and karaoke during a long fieldwork stay in Japan.

As I moved on to a postdoc position, this book project was on hold for a while as there always seemed to be another matter to attend to. I owe gratitude to Miranda and Greg for ceaselessly reminding me to pick up the manuscript again and for being there as mentors. Although he never received official acknowledgment for his efforts as my third supervisor, Greg's willingness to read draft sections and to give frank and thought-provoking feedback has helped greatly in further developing this manuscript. More generally, I would like to express my gratitude to those more experienced colleagues, visiting professors, and other researchers who took the time to give their feedback on various aspects of this project during and outside of conferences and workshops. Two people who cannot go unmentioned for their invaluable wisdom and willingness to share it are Richard Samuels and Jacques Hymans. Of course, all of this would have not been possible without the many interview partners and experts in Japan, who were generous with their time and kindly assisted with filling in the blanks, fact checking, and updates.

This research has been funded by the German Research Foundation (DFG); the German Academic Exchange Service (DAAD); the European Union Program for Education, Youth and Sports (Erasmus+); and the Technical University Munich (TUM). During different research stays in Japan, I have been warmly welcomed and supported by the Sophia University Graduate School of Global Environmental Studies, the Tokyo University Institute of Social Science (ISS), the German Institute for Japanese Studies in Tokyo (DIJ), and the Kyoto University Graduate School of Global Environmental Studies as well as the Center for Advanced Policy Studies. When it came time to finish up and polish the manuscript, I would like to thank the two anonymous reviewers for helping to improve the manuscript with their comments, the Social Science Japan Journal for permission to reprint a previously published table, and the editor Sarah Grossman at Cornell University Press for her guidance throughout the process.

Last, but not least, I would like to thank my family and friends, especially the "usual suspects," for bearing with me and for expecting nothing less than completion from me. I dedicate this book to my mother and my late grandfather.

Japanese names are presented in the traditional Japanese order with the family name stated first.

JAPAN'S NUCLEAR DISASTER AND THE POLITICS OF SAFETY GOVERNANCE

Introduction

JAPAN'S NUCLEAR DISASTER AND REGULATORY POLITICS

Images of the earthquake, tsunami, and nuclear accident that assailed Fukushima, Japan, on March 11, 2011 (commonly known in Japan as "3.11") were broadcast to millions of televisions around the world. Many people watched live as the events unfolded over days, evolving into partial core meltdowns in three of six reactors. In a sequence that became world famous, Unit 1 blew up on March 12, followed by a cloud of white smoke, leaving a skeleton where the reactor building had stood. Japan reported the nuclear accident to the International Atomic Energy Agency (IAEA), as required. On the day it occurred, Japan's Nuclear and Industrial Safety Agency (NISA) rated it as a level 4 out of seven levels on the IAEA scale: "accident with local consequences" (IAEA 2011b). In the following days, nuclear safety experts around the world analyzed the camera footage and other available data to gauge the severity of the accident. While the plant operator Tokyo Electric Power Company (TEPCO) and the Japanese government reassured the public that no meltdown had taken place yet, French nuclear safety authorities questioned the validity of the level 4 classification (BBC 2011) and nuclear experts in the United Kingdom warned of nuclear meltdowns as early as March 14 (Guardian 2011a). One month after 3.11, NISA revised its assessment to a level 7 "major accident" (IAEA 2011a), the highest level on the IAEA accident scale. TEPCO has since admitted that it should have declared a nuclear meltdown much sooner. Second in severity only to Chernobyl in 1986, a disaster of such magnitude in high-tech Japan raised urgent questions about how to govern nuclear safety worldwide.

In 2011, IAEA Director General Yukiya Amano stated that the Japanese nuclear disaster "caused deep public anxiety throughout the world and damaged

1

confidence in nuclear power" (UPI 2020). In response, the German government, facing mass public demonstrations and a rapid shift in public opinion against nuclear power (Infratest 2011), appointed an "ethics commission for safe electricity generation" to assess the overall risks and benefits of nuclear power. Comprising diverse societal stakeholders, ranging from academics to church representatives, the commission advised against reliance on nuclear power due to incalculable long-term risks associated with accidents and radioactive waste storage (Ethik-Kommission Sichere Energieversorgung 2011). Following this recommendation, German Chancellor Angela Merkel announced that Germany would phase out nuclear power by 2022. Belgium embarked on a similar stepwise phaseout of nuclear power by a set date in the 2020s. Spain and Switzerland foreclosed the construction of additional nuclear power plants. Other nations, such as the United States, China, the United Kingdom, and France, retained nuclear power as part of their energy policy. Responses to the trust-shattering nuclear accident varied around the world and, to the surprise of many, Japan was one of the countries that decided to retain nuclear power.

However, 3.11 exposed severe deficiencies in Japan's nuclear safety governance. To begin with, Japan's Nuclear and Industrial Safety Agency (NISA) failed to prevent the Fukushima nuclear disaster. Shortly before 3.11, it had approved the oldest of the reactors at the Fukushima Daiichi nuclear power plant as safe for another ten years of operation. The inadequacy of safety governance institutions in Japan was further highlighted by NISA's failure to correctly assess the severity of the accident, evident from its initial rating as level 4 on the IAEA scale of what turned out to be a massive level 7 nuclear meltdown. Moreover, NISA lacked expertise to take on its predetermined role as a government adviser in crisis management (Kushida 2016), forcing then Japanese Prime Minister Kan Naoto and his government to look elsewhere for expert advice (Kan 2012). The nuclear accident cast a harsh spotlight on nuclear safety governance flaws, a rude awakening for a country that had thought it enjoyed high levels of nuclear safety.

In fact, an accident investigation committee created by the Japanese Diet (the Japanese parliament) exposed deep-seated flaws in how Japan had hitherto regulated nuclear safety. The Nuclear Accident Independent Investigation Committee (NAIIC), the first of its kind in Japan's postwar history, was equipped with far-reaching competencies to view otherwise confidential documents and to conduct public hearings with persons of interest. It came to a shattering conclusion about Japan's nuclear safety governance:

> The TEPCO Fukushima Nuclear Power Plant accident was the result of collusion between the government, the regulators and TEPCO, and the lack of governance by said parties (NAIIC 2012, 16). The regulators did

not monitor or supervise nuclear safety. The lack of expertise resulted in "regulatory capture," and the postponement of the implementation of relevant regulations. They [regulators] avoided their direct responsibilities by letting operators apply regulations on a voluntary basis. Their independence from the political arena, the ministries promoting nuclear energy, and the operators was a mockery. (NAIIC 2012, 20)

The investigation concluded that the accident was not caused by a natural disaster per se, but rather by human failure to address known risks related to natural disasters.

To address these governance flaws, in September 2012 the Nuclear Regulation Authority (NRA) was created as a so-called independent regulatory commission. The new agency unified different functions related to the safety of nuclear power plants, safeguarding nuclear materials and nuclear security. Looking at the safety governance of nuclear power plants, these reforms finally separated nuclear regulation functions from nuclear power promotion, strengthened public accountability, and imposed rigorous and far-reaching safety regulation vis-à-vis the nuclear industry. Hence, instead of phasing out nuclear power in response to 3.11, Japan created an independent nuclear safety regulator in order to reinstill domestic and international trust in nuclear safety and to enable the continued operation of Japan's nuclear power plants.

Contrary to low expectations (Aldrich 2014; Cotton 2014; Vivoda and Graetz 2014; Kingston 2014; Hymans 2015), the NRA defied pronuclear actors' pressure for a quick and low-cost return to nuclear power. The LDP (Liberal Democratic Party) government made its wishes to reduce some of the new safety requirements clear after 2012, but it did not get satisfaction from the NRA. This book aims to explain the puzzling existence of an independent nuclear safety regulator in Japan which did not backslide on the more stringent safety measures introduced, even after it became clear that those measures were going to force numerous reactors into decommissioning.

3.11 Shook Japan's Nuclear Policy to Its Core

As accidents often do (Birkland 2007; Balleisen et al. 2017), the nuclear meltdowns and revelations about nuclear safety governance flaws led to an intense debate about reforming Japan's nuclear power policy. Japanese Prime Minister Kan, head of a government formed by the former opposition party, the Democratic Party of Japan (DPJ), championed an unprecedented call for a nuclear phaseout to achieve "a society that is not dependent on nuclear power" (Prime Minister's Office 2011). At the same time, Japan saw the largest public demonstrations in decades, which reached their zenith in September 2011, with a rally of sixty thousand people at Meiji Park in Tokyo, including Nobel Prize winner Ōe Kenzaburo (Hasegawa 2014). In contrast, the old pronuclear power elites, such as the LDP, the Ministry of Economy, Trade, and Industry (METI), Japan's leading industry association Keidanren, and nuclear power plant operators, such as TEPCO, sought to return to pre-3.11 levels of nuclear power generation as swiftly as possible.

Next to calls for radical policy change, such as phasing out nuclear power altogether, the post-3.11 debate about the future of nuclear power revolved around improving nuclear safety. Prime Minister Kan initiated a process of restructuring Japan's nuclear safety institutions and ordered an immediate "stress test" to review the safety of Japan's nuclear power plants, following the model set by the European Union (EU). After 3.11, the EU embarked on "comprehensive risk and safety assessments ("stress tests") of nuclear power plants" (EU Commission 2011).[1] In Japan, the stress test was conducted by NISA, the agency that had failed to prevent the accident in the first place, misjudged its severity, and proven unable to take on its assigned crisis management role. When the first nuclear power plants in Ōi passed the stress test and were restarted in the summer of 2012, networks of old and new antinuclear civil society organizations mobilized up to two hundred thousand people protesting in front of the Diet, the Japanese parliament (Wiemann 2018). In an August 2012 public opinion poll, 69 percent of the Japanese public regarded nuclear power as either very, somewhat, or slightly dangerous (Shibata and Tomokiyo 2014, 47).

Calls for changes in Japan's nuclear safety governance not only came from the Kan government but were strongly supported by international actors such as the IAEA. Immediately following 3.11, the IAEA began a process of revising global nuclear safety guidelines. It culminated in the 2016 IAEA *Safety Standards for People and the Environment*, which came to include "freedom from political pressure" (IAEA 2016a) as a lesson learned from Japan's pre-3.11 safety governance failure. During the reform process in Japan, IAEA representatives backed the idea of a stronger, more independent nuclear safety agency. Huge numbers of concerned citizens protesting in front of the parliament also lent strong support to the reform process. As the largest opposition party at the time, the LDP held the necessary seats in the Diet to block safety governance reforms, but instead it opted to use its power to push for an independent safety administration with a strong legal framework. As a result of cooperation and compromise between the DPJ government and the LDP opposition, the Diet passed the Nuclear Regulation Authority Establishment Act in June 2012.

Soon after, the LDP won the December 2012 general elections with a landslide victory despite its stance being "the most positive of any party and thus the furthest from the majority of an electorate soured on nuclear power" (Endo, Pekkanen, and Reed 2013, 60). The reelection of the LDP and pronuclear Prime Minister Abe Shinzō put an end to calls for a complete nuclear phaseout. Soon after, Abe announced that "over the course of roughly three years we will assess the futures of existing nuclear power plants and transition to a new stable energy mix over ten years" (Abe Shinzo 2013). Ahead of the 2014 snap elections, the Abe government adopted a strategy to depoliticize nuclear power with a "shifting of responsibility for decisions about the safety of nuclear restarts to the newly created Nuclear Regulation Authority (NRA)" (Hughes 2015, 203). Strikingly, the LDP has won one national election after another despite the unpopularity of Abe's pronuclear stance.

Around the time the LDP returned to government, large-scale antinuclear protest subsided. While the "anti-nuclear movement wave" (Wiemann 2018) marked the end of Japan's "ice age of social movements" (Pekkanen 2006), it did not translate into a sustained antinuclear movement. As part of the "new protest cycle" that emerged after 3.11 (Chiavacci and Obinger 2018), attention turned away from nuclear power and towards the Abe government's national security policy agenda. Demonstrations focused on protesting against the State Secrecy Law (2013), the Security Laws (2015), and the Conspiracy Law (2017). What remained from the large antinuclear demonstrations of 2011 and 2012 was a small group of people protesting against nuclear power at a street corner in front of the Prime Minister's Office every Friday.

The turn of the year from 2012 to 2013 seemingly marked a reversion back to nuclear politics as usual: a pronuclear government in power and able to win election after election, no large public demonstrations against nuclear power, and a new safety agency in place to ensure a smooth return to nuclear power with better safety standards as a means to regain international and domestic trust lost after 3.11.

Why Did Japan Hold on to Nuclear Power Despite the Obvious Risks?

As LDP Prime Minister Abe explained his government's stance: "Our resource-poor country cannot do without nuclear power to secure the stability of energy supply while considering what makes economic sense and the issue of climate change" (Japan Today 2016). Concretely, the Long-Term Energy and Supply Outlook, adopted by the government in April 2015, set a target for nuclear power to account for 20 to 22 percent of electricity generation in 2030 in order to achieve the triad of energy policy goals: energy security, environmental and climate friendliness, and economic efficiency (known as "3E"). A return to a strong nuclear policy was considered pivotal to Abe's economic policies to reignite growth, called Abenomics (Kingston 2016; Incerti and Lipscy 2018); to Japan's efforts to raise its energy self-sufficiency rate, which reached a low point after 3.11 (Vivoda 2014); and to lowering greenhouse gas emissions in line with international commitments under the United Nations climate regime (Kameyama 2019). Hence, Japan's "power elite, and notably the so-called 'nuclear village' of big business, the electrical utilities, and key government ministries, wanted to return to business as usual" (Hymans 2015, 113).

The so-called nuclear village,[2] a term mainly used by critics to convey the intimate proximity of relations, was centered on METI, the nuclear industry—plant manufacturers and regional electric utilities—and the LDP.[3] It resembled a classic "iron triangle" of policymaking and implementation. The nuclear iron triangle was dominated by vested interests, had a propensity for making agreements behind closed doors, and promoted a nuclear "safety myth" about the absolute safety of Japanese nuclear plants (Cotton 2014; Vivoda and Graetz 2014; Kingston 2014). According to the NAIIC, Japan's pre-3.11 nuclear safety governance setup, designed in the immediate postwar period, was influenced by pronuclear politicians and ministries, and depended on industry expertise, allowing the industry to "capture" regulation and twist it in its favor at the expense of public safety. This regulatory governance design exhibited path dependency for half a century, despite multiple reform attempts in the aftermath of smaller nuclear accidents in Japan.

Japan's Game-Changing Post-3.11 Nuclear Safety Reforms

Many studies of nuclear energy in post-3.11-Japan have drawn on the notion of crisis-induced change to ask whether it brought about fundamental changes. Early studies failed to anticipate the crucial changes resulting from the creation of the NRA (Samuels 2013; Elliott 2013; Al-Badri and Berends 2013). The new agency took over staff from the previous captured safety administration, a pronuclear chairman headed the new agency, and most observers expected significant pressure from powerful pronuclear actors on the NRA (Aldrich 2014; Cotton 2014; Vivoda and Graetz 2014; Kingston 2014; Hymans 2015). Some scholars attested to the NRA's strict stance on regulation and its apparent ability to resist political pressure, but they omitted an analysis of the reasons behind it and expressed doubts about its ability to remain independent in the longer run (Aldrich 2014; Hymans 2015; Kingston 2014; Shiroyama 2015).

Defying low expectations, the new nuclear safety agency conducted thorough and time-consuming safety checks. By early 2022, the NRA had completed seventeen safety reviews—a far cry from the review of all reactors demanded by Abe by 2016. In the process, the NRA imposed massive safety investments on the nuclear industry, especially electric utilities operating nuclear power plants. According to the team leader of the IAEA Integrated Regulatory Review Service Mission to Japan: "In the few years since its establishment, the NRA has demonstrated its independence and transparency. It has established new regulatory requirements for nuclear installations and reviewed the first restart applications by utilities" (IAEA 2016c, 1). There is a gap in understanding how Japan's independent nuclear regulator was able to *persist* in the face of a pronuclear iron triangle pushing for a return to pre-3.11 business as usual.

This book argues that the answer to this puzzle lies in the initial steps taken soon after the accident. In fact, reformers and the first NRA board were keenly aware of the need to break with a path-dependent system of regulatory capture and put in place an institution championing independent decision-making, more stringent safety standards, transparency, and in-house training programs to raise expertise levels. The chairman some had written off as part of the nuclear village turned out to be an enthusiastic reformer who quickly developed a political savviness few expected from a scientist. Defending its independence, the NRA issued operating permits without succumbing to political and industry pressure to speed up the process. Binding safety standards shifted power from the nuclear industry to the safety agency and ended the practice of partial industry self-regulation. Unprecedented levels of transparency, aimed to regain trust in Japan's safety governance, informed citizens who were previously excluded from nuclear politics and invited international scrutiny of reform outcomes. With the creation of an independent and transparent safety agency raising the costs of technical compliance and social acceptance, Japan's nuclear safety governance abruptly shifted from aggressive promotion to rigorous oversight.

This change is interesting in light of global nuclear safety regulation trends. Since the 1980s, independent regulatory agencies (IRA) have begun to spread around the world (Jordana, Levi-Faur, and i Marín 2011). In a process dubbed "diffusion," there has been a process of homogenization where organizations with similar functions but located in different countries take on a similar shape (DiMaggio and Powell 1983). The significant increase in the number of independent regulatory agencies, especially in the United States and in Europe, made profound changes to the political economy of many countries by turning them into "regulatory states" (Levi-Faur 2011). At the same time, independent nuclear safety governance has been promoted by the International Atomic Energy Agency (IAEA) in general (IAEA 2003, 2006) and for Japan in particular (IAEA 2007) as well as by people such as Richard Meserve (Meserve 2009), former head of the US Nuclear Regulatory Commission (NRC), which is often touted as a model for an independent nuclear safety regulator. Examples of nuclear safety agencies whose independence has increased in the 2000s include those in Spain (Bianculli, Jordana, and Juanatey 2017) and France (Arnhold 2021). However, the diffusion of IRA has mostly evaded Japan, with the exception being the "[Japan] Fair Trade Commission (JFTC) as a rare independent regulator in Japan" (Mogaki 2019, 17). Given the general absence of independent regulatory agencies, and the collusion that facilitated the 2011 Fukushima nuclear accident, Japan's new independent nuclear safety agency is a case study worth paying closer attention to.

This nuclear safety governance shift is of practical relevance, too, because Japan is an important player in international energy and environment issues. It is the world's third biggest economy as measured by nominal GDP and the fourth biggest in purchasing power parity. Japan's economy is highly energy dependent: the country had the fourth biggest electricity consumption worldwide in 2018. How Japan generates this massive amount of electricity matters for its economy, its energy security, and for the environment. Japan currently stands as the fifth biggest emitter of greenhouse gases worldwide. Since the early postwar period, Japan has promoted nuclear power as a technology which promised endless clean electricity to support the postwar economic recovery and to make the country less dependent on energy imports. Nor was the influence of the Japanese nuclear power complex limited to the domestic economy. As part of the "nuclear renaissance" in the 2000s, Japanese manufacturing companies Toshiba, Mitsubishi, and Hitachi moved into the heart of the global nuclear industry: Toshiba increased its stakes in the nuclear power plant manufacturer Westinghouse, while Hitachi and Mitsubishi entered tie-ups with GE and France's Areva, respectively. By 2011, Japan had the third largest number of nuclear reactors for commercial electricity generation, after the United States and France. Given its dominant position in the global nuclear industry, how Japan regulates nuclear power has practical implications for global energy and nuclear power development as well as efforts to curb greenhouse gas emissions.

Regarding a possible lasting impact of 3.11 on Japanese nuclear politics, Samuels (2013, 120) concluded: "It is too early to tell and too soon to conclude otherwise." The LDP's return to government seemed to signal Japan's return to pre-3.11 nuclear politics. Given nuclear power's pivotal position for Japan's economy, energy security, and climate impact as well as the support from powerful political, administrative, and industry actors, a return to heavy reliance on nuclear power seemed only a matter of time. However, a decade after 3.11, the future of nuclear power in Japan remained up in the air.

The NRA was instrumental to this development by breaking with decades-old safety governance practices. It created a hurdle for nuclear reactor restarts by imposing strict safety measures. Due to thorough and time-consuming safety checks, Japan's return to nuclear power was a sluggish one. According to IEA data, in 2018–19 nuclear power provided only 6 percent of Japan's electricity, and this figure even declined to about 4 percent in 2020. Meanwhile, the costs of refurbishing reactors to meet the stringent new safety requirements have skyrocketed from an estimated one trillion yen (roughly USD 10 billion) in January 2013 to five trillion yen (a little less than USD 50 billion) in July 2019 (Asahi Shimbun 2019a). From 2011 to 2021, the number of nuclear reactors available for electricity generation dropped from fifty-four to thirty-five (JAIF 2022).

Meanwhile, electric utilities, which operate nuclear power plants, applied for safety checks to restart twenty-seven of the remaining reactors. Citizen groups have used legal means to challenge decisions to restart reactors, resulting in more safety-related class-action lawsuits brought against nuclear power plants between 2011 and 2015 than in the preceding four decades (CNIC 2020). Experts estimate that nuclear power will provide at most 15 percent of Japan's electricity by 2030 (Koppenborg 2016), and perhaps as little as 10 percent (Izadi-Najafabadi 2015). Apparent administrative and political support by powerful political actors has not translated into swift policy implementation. Rather, nuclear safety governance reforms are at the core of a shift in Japan's nuclear politics that curtailed pronuclear actors' power to implement nuclear power policy (Koppenborg 2021). In fact, a new safety agency, created to enable a continued use of nuclear power, has ushered in a new path in safety governance and, by extension, nuclear power politics.

Aim of the Book

The aim of this book is to explain the puzzling existence of an independent nuclear safety regulator in Japan, which did not backslide on the more stringent safety measures introduced despite the government making it clear after 2012 that it wishes to reduce some of these requirements. I reconstruct the evolution of Japan's nuclear safety governance and nuclear politics since the early postwar period, which encompasses half a century of path dependence as well as fundamental changes following 3.11. For more than half a century, Japan's powerful nuclear iron triangle fostered a captured safety governance system that largely kept critics and the public out of nuclear politics. The 3.11 events, however, brought the legitimacy of this system into question and brought critics of "regulatory capture" to the forefront. While pronuclear actors defended the nuclear policy option, actors eager to bring about change in nuclear safety governance—concerned politicians, experts, citizens, and international actors—successfully pushed for fundamental institutional changes. This contrast between industry-led safety regulation pre-3.11 and independent and transparent safety governance is extremely apparent in the case of Japan.

The process of reforming safety governance institutions after 3.11 is at the heart of this study. An analysis of safety governance reforms draws attention to the ways in which safety governance institutions affect politics and vice versa. It highlights how the design of safety institutions can become a political game changer by redefining the relationship between state, industry, and public. It sheds light on the role of knowledge, both as technical expertise and as public knowledge of decisions made, showing how "knowledge is power."[4] Industry-led opaque safety governance occurs when the industry produces expertise for safety governance, while the public is kept largely in the dark about the risks involved. In contrast, safety governance rooted in independent expertise and public participation provides the basis for the state to regulate risks in the public interest. Thus, power relations in such a policy domain are shaped by whether the state has the necessary expertise to regulate risks as opposed to leaving it to the industry and whether information about this process is available to the public. This book thus stresses the role of expertise and transparency and contributes to understanding how they can shift the power distribution among state, industry, and public actors. Design choices for safety governance institutions during times of reform shape the knowledge and information landscape in the political process, redistributing power among different actors and shaping political conflicts.

Many new technologies share with nuclear power the promise to create a better future but also the potential for massive negative consequences. Nuclear power technology is emblematic of modern "risk society" and the double-edged sword of technological progress, where "the social production of *wealth* is systematically accompanied by the social production of *risk*" (Beck 1992, 19). The same applies to other technologies that promise an increase in wealth and prosperity, such as smart green technologies, artificial intelligence, and related digital technologies. Such "high-risk high-return" technologies pose challenges for political decision makers similar to those Japan faced in the 1950s when it set up nuclear safety governance. This is certainly true in democratic countries where technology choices follow market mechanisms, and public opinion and participation matter; similar challenges may even prove to exert some influence in authoritarian countries.

Hence, the case of Japan provides important lessons for safety and the distribution of benefits and—potentially global and as yet unforeseeable—risks. It highlights the importance of expertise in state agencies and informational transparency as a door opener for a societal debate about acceptable risks. Facing a multitude of emerging technologies, it is in the interests of states to procure the necessary knowledge and establish a legal framework to avoid falling prey to one-sided industry influence and possibly sacrificing public safety for the promise of technological solutions that have not yet been thoroughly vetted. This book seeks to deepen our understanding of the political nature of safety governance institutions and offers a dynamic account of how such institutions are created and, in turn, influence levels of public accountability and industry influence over the risks with which the public is made to live.

Regulatory Politics: Nuclear Politics and Safety Governance

There is nothing apolitical about regulation. This book explores this particular aspect exemplified by Japan's nuclear safety reforms: that safety governance institutions and politics are not separate but inextricably intertwined. A regulatory governance setup predetermines whose knowledge is considered, whose voice is heard in the debate surrounding the safe use of high-risk high-return technologies, and who holds power in the respective public policy conflict. Hence, this book stresses the relationship between power, knowledge, and public policy conflict. To highlight the link between regulation and politics, this book uses the term regulatory politics.

Regulatory politics refers to a political outcome, where safety governance shapes the politics surrounding a technology. Safety governance institutions structure the distribution of power among industry, politicians, administration, and the public, with important implications for deciding and implementing political responses to technological risks. Transparent and independent safety governance altered Japan's nuclear politics by redistributing power. The NRA curtailed industry power to influence safety standards, which shifted power to determine risks for society back to the state and significantly raised the technical costs of nuclear power. Transparency and expanded safety standards galvanized public participation, often in the form of legal action against restarts of nuclear reactors, which added further social acceptance costs—and protected the new agency from overt attempts to undermine its independence. In Japan, seemingly unremarkable nuclear safety administration reforms quickly turned into a game changer for nuclear politics, particularly policy implementation.

At the same time, regulatory politics describes the political process of designing regulatory governance institutions. When an extant safety governance design is undermined by an accident, political decision makers have an opportunity to reconfigure governance institutions to shift the balance between promoting industry interests and expanding public accountability and participation. Regulatory politics as a design process thus creates an institutional structure that shapes the politics surrounding the respective technology for years to come. In this case, it refers to reforms that dismantled the old institutions and practices in the aftermath of 3.11. Under conditions of heightened domestic and international attention after 3.11, actors eager to improve nuclear safety governance pushed for reforms. They overcame resistance from vested pronuclear interests and designed fundamentally different governance institutions that stressed independence and transparency.

Regulatory politics combines a regulation theory view of regulatory governance institutions with a historical institutionalist view of institutional change. First, regulation theory has seen a long-standing debate about the motivations of regulation that developed two distinct approaches. The "public interest" theory states that governments impose regulation to correct market failures and to thereby benefit consumers and raise social welfare. In contrast, the "private interest" thesis, also known as "regulatory capture," posits that the industry will come to influence the regulator that was supposed to subject it to regulation to advance public interest goals. Somewhat problematically, this debate pins public and private interest against one another: captured regulation that benefits the industry at public expense versus regulation in the public interest (Priest 1993). In reality, the distinction is not always so clear. Take electric utilities in Japan, for example: the goals they pursued, such as supplying affordable, clean, and stable electricity to fuel economic growth without polluting the environment and fueling climate change, were indeed in the interest of the Japanese public.

Rather than the motivations of regulation, the focus here is on how governance institutions shape the relationship between state, industry, and public. I here borrow insights from the concept of regulatory capture and its view of industry influence over regulation (Wilson 1980; Mitnick 2011) and from studies of the "regulatory state" that focus on the rise of independent regulatory agencies established to further public-interest regulation (Levi-Faur 2011; Jordana, Levi-Faur, and Fernández-i-Marín 2011). Regulation theory highlights that regulation is a conflictual process that will determine where the balance between public and private interest falls between the two extremes of private-interest and public-interest regulation. In a risk society, regulation not only distributes the socially produced risks that go hand in hand with the technology-based production of wealth but depending on the governance mode also influences who has a say in this crucial distributive process.

Second, the historical institutionalist view of institutional change rests on the central theorem that institutions exhibit stability over long periods of time, called path dependence, but experience fundamental agency-driven change during so-called critical junctures (Steinmo, Thelen, and Longstreth 1992; Hall and Taylor 1996; Thelen 1999; Pierson 2000a; Capoccia and Kelemen 2007; Capoccia 2016a). Historical institutionalism is an established research tradition in domestic and comparative politics, but the renegotiation of institutional design following focusing events, such as accidents or economic crises, also has attracted increasing attention from students of policy and regulation change (Balleisen et al. 2017; Birkland 2007; Baumgartner and Jones 1993; Kingdon 1984; Sabatier and Weible 2007). Drawing on regulation theory and theoretical insights into critical junctures, I reconstruct the politics of reforming a regulatory organization, the institutionalization of a different governance mode, and the constraints it created for previously powerful actors. This contributes to attempts to grasp the political dimensions of institutional change by accounting for regulatory governance reforms as a way to (re)shape politics by (re)distributing power between state, industry, and public in the aftermath of a focusing event.

Institutional Structure

Institutions range from a constitutional order to state-business relations to the policies and regulations governing interactions in a policy domain. Institutions and organizations are clearly linked in political practice, but they need to be distinguished for analytical clarity. Institutions are the humanly devised "rules of the game" that structure interactions of different organizations (North 1990). These rules encompass "the formal and informal procedures, routines, norms and conventions embedded in the organizational structure of the polity or political economy" (Hall and Taylor 1996, 938). Hence, it is important to study both collective actors (organizations) and the formal and informal rules (institutions) that structure their political interaction.

By distributing power, institutions determine the basic nature of politics. Historical institutionalism adopts Schattschneider's (1961) notion that conflict is at the heart of politics and that institutions organize some interests into politics while others are organized out. The "rules of the game" embedded in these institutions distribute power among different actors (Hall and Taylor 1996; Thelen 1999). Power has many faces. First, agenda-setting power is the ability to put an issue on the political agenda or keep it out of the spotlight. Second, decision-making power refers to the ability to decide for oneself or to prevent a decision from being taken at all. Third, (de)politicization power, or domination power, refers to the ability to convince others to accept one's own interest as their own and to refrain from attempting to put a topic on the political agenda in the first place (Lukes 1974, 24–25). Institutions, and the organizations they are embedded in, define actors' access to political agenda setting and decision making. What is missing in this conception and requires more attention is the distribution of power in policy implementation, because decisions that are not implemented remain political paper tigers.

Understanding institutional change, and the resulting change in power distribution, requires a thorough analysis of preceding path dependence (Thelen 1999). A state of stability is the result of feedback mechanisms involving increasing returns and vested interests. Feedback mechanisms describe the process whereby actors adapt their behavior to a given institutional arrangement in ways that reflect it and reinforce the underlying logic of the system. Increasing returns frequently occur because the costs of switching to a previously plausible alternative increase with each step down a chosen pathway. The attractiveness of the current path compared to other previously possible options increases over time, decreasing the likelihood of completely changing paths (Pierson 2000a). The distributional nature of institutions empowers some actors while marginalizing others, creating winners and losers within a particular institutional order. Empowered actors will be particularly vested in the institutional arrangements and have an incentive to sustain the arrangement and their position. Those who are kept out have an incentive to seize an opportunity to challenge it (Thelen 1999).

Within a given policy domain, for example nuclear politics, the distributional nature of institutions translates into rather stable actor constellations, in which some enjoy access to decision making while others are excluded. Those in decision-making positions, the winners of an institutional order, constitute a "policy monopoly" (Baumgartner and Jones 1993) or "dominant coalition." Usually, a policy domain consists of at least one dominant and one minor coalition (Sabatier and Weible 2007). Japan's policymaking process is commonly described as centered on an iron triangle or elite triumvirate of the long-term ruling LDP, the bureaucracy, and big business. Hence, the iron triangle is a feature common to many dominant policy coalitions in Japan. There is an ongoing debate about which of these actors in the dominant coalition governs Japan—the bureaucracy, politicians, or the private sector.

The characterization of Japan as a "developmental state," in which the Ministry of International Trade and Industry (MITI, now METI) effectively orchestrated Japan's postwar economic recovery using "administrative guidance" (Johnson 1982, 1995) is possibly the most famous example of the argument that the Japanese elite state bureaucracy dominates policy decision making (Pempel

1982; Vogel 1999). Other scholars stressed the role of politicians as "principals" and the bureaucracy as their "agents," which puts politicians in the driving seat of Japanese politics (Ramseyer and Rosenbluth 1993). Following administrative restructuring as part of the Hashimoto reforms in the late 1990s and early 2000s, which strengthened the previously weak Prime Minister's Office, some saw the potential for a "presidentialization" of the policymaking system (Poguntke and Webb 2005). Yet another perspective stresses that business interests win, at least during times of "quiet politics," when the public pays little attention to an issue and there is low voting salience (Culpepper 2010).

The nuclear power policy domain exhibited particular power relations among those in the iron triangle. While the Japanese parliament holds both executive and legislative power, in practice "once government approval, most of the time LDP approval, had been given, the implementation [of nuclear power policy] was left up to MITI in the form of granting licenses" (Cohen, McCubbins, and Rosenbluth 1995, 195). Theoretically, this put MITI in a powerful position. However, Johnson, the famous advocate of the concept of administrative guidance by MITI, concluded that electric utilities were strongly opposed to government controls and "developed differently from industries that have been directly fostered by the government" (Johnson 1978, 132). The relationship has otherwise been described as one of "reciprocal consent," where the bureaucracy "achieves jurisdiction and is denied control. . . . The Japanese bureaucracy does not dominate, it negotiates" (Samuels 1987, 260). With political blessing from the LDP, electric utilities occupied a strong position in nuclear energy policymaking, and especially in the implementation process.

To sustain their position, it is in the interest of the dominant coalition 1) to shape the rules of access in such a way to discourage marginalized actors from participating and 2) to associate a positive image with the political issue to convince marginalized actors that decisions are made in their own interest, eliminating the need to seek participation in decision-making (Baumgartner and Jones 1993). Shaping the rules of access serves to retain the first two faces of power, agenda-setting and decision-making power, while associating a positive image helps to sustain domination or depoliticization power. It is difficult to overcome the separation between dominant coalition and outsiders, because both the institutional structure of the policymaking process, which determines rules of access, and the images (i.e., ideas and understandings) people associate with a political issue exhibit continuity over long periods (Sabatier and Weible 2007). For the winners of an institutional order, limiting the number of actors by keeping extant institutions and supporting incumbent policy ideas is an important strategy in politics.

The principal-agent relationship that put politicians in the driver's seat in Japan included judicial manipulation (Ramseyer and Rosenbluth 1993), which refers to the government using the appointment of judges as a way to punish those who ruled against the LDP's policy stance, at least on high-profile partisan disputes, for example the status of Japan's military forces (Ramseyer 2019). While courts presented an avenue for citizens pushing for stronger pollution regulation (Upham 1987), none of the lawsuits seeking to halt nuclear power development before 2011 were successful (Kingston 2014, 111; interview with H. Ban, 2018). This exemplifies the dual role of the Japanese judiciary—it can either function as an avenue for citizens to engage in politics or as a gatekeeper for politicians by keeping them out.

Japan's civil society was comparatively weak since the "laws governing the non-profit sector made it extremely difficult for groups to obtain non-profit status" (Schreurs 2003, 244). Robert Pekkanen details how small local groups "such as neighborhood associations are promoted by the government" while "it is hard for autonomous groups to become large and hard for large groups to be autonomous" (Pekkanen 2003, 133). Consequently, antinuclear protests were limited to NIMBY (not-in-my-backyard) protests against the development of nuclear power plants. Few in numbers at first, a "new wave of anti-nuclear protests" emerged after the 1986 Chernobyl accident (Honda 2005; Hasegawa 2004). These NIMBY protests were network based and characterized by their rejection of bureaucratic organizational structures (Hasegawa 2004). In response to—sometimes successful—NIMBY protests, the Japanese state displayed a "Machiavellian" stance by seeking to shape social preferences to serve its policy choices and by avoiding areas with stronger potential for resistance in the process of developing nuclear power (Aldrich 2010).

In Japan's nuclear energy policy domain, pronuclear actors successfully limited public participation, shaped public opinion, and devised ways around local protest in order to develop Japan's nuclear power program. That was the case, at least, until 3.11. The reconstruction of pre- and post-3.11 nuclear safety governance emphasizes that regulatory reforms, prompted by a focusing event, constituted a critical juncture and ushered in a new path because they undermined the feedback mechanisms sustaining the power structure of the previous institutional arrangement. Hence, governance reforms can become political game changers if they weaken the position of powerful actors that sustained the previous path.

Institutional Design

To understand the redistribution of power between the state, nuclear industry, and the public in Japan, this book analyses the process of designing safety governance institutions in the aftermath of 3.11. At the time, the LDP was not in government and the Democratic Party of Japan (DPJ), a political party with fewer ties to the old power elite, took over the reform rudder. Nuclear safety governance reforms championed by the DPJ, and supported by the LDP while in opposition, institutionalized a new power distribution, thus constraining and enabling different actors than before.

A window of opportunity for change opens up when the credibility of existing institutional arrangements is challenged. This is often the result of shock events, such as accidents, disasters, and economic crises, which throw a spotlight on deficiencies of the existing institutional order. Shock events have alternatively been called external shocks or perturbations (Kingdon 1984; Baumgartner and Jones 1993; Blyth 2002; Hogan and Doyle 2007; Birkland 2007; Capoccia and Kelemen 2007; Balleisen et al. 2017). Such a shock event does not automatically translate into change. Some attempts at change result in the previous arrangements reasserting themselves, an outcome termed a "near miss" (Capoccia and Kelemen 2007). This is exemplified by different responses to the nuclear accident around the world, ranging from reconfirming existing safety mechanisms to phasing out nuclear power altogether. In Japan, it induced calls for a complete nuclear phaseout, which failed to materialize. Thus, a shock event opens up a window of opportunity for major institutional change but is not sufficient in itself.

Understanding the process whereby actors bring about change during a window of opportunity to design institutions is at the heart of ongoing debates in historical institutionalism. The strand of literature that seeks to better understand rapid and fundamental changes that mark the beginning of a developmental pathway studies so-called critical junctures. These "crucial founding moments of institutional formation" (Thelen 1999, 387) are brief phases of institutional fluidity, in which actors have an opportunity to create new institutional arrangements (Capoccia and Kelemen 2007). Rather than incremental change based on opportunities and limits for agency within the constraints of institutions (Thelen 1999; Mahoney and Thelen 2010; Capoccia 2016b; Streeck and Thelen 2005), studies of critical junctures focus on shock-induced institutional change and why actors opt for a certain institutional design while disregarding others (Capoccia and Kelemen 2007; Hogan and Doyle 2007). As crises are a common occurrence in modern risk society, where technological progress produces new risks (Beck 1992, 19), it is important to understand the dynamics of crises and change and how they translate into a new institutional arrangement.

Explanations of why a window of opportunity turns into a critical juncture, as was the case with Japan's nuclear safety governance reforms, focus on ideas and power as change mechanisms (Capoccia 2016a; Béland, Carstensen, and Seabrooke 2016; Hogan 2019). Both include the notion that previously uninvolved actors become policy entrepreneurs who push for change. There are three categories of agents of change: First, policy entrepreneurs produce policy-related ideas; they include bureaucrats, academics, and interest groups who have access to the political process. Second, political entrepreneurs—who are mostly politicians and often opposition leaders—are figureheads who introduce a new idea into the political process. Third, outside influences are domestic and international actors that are located outside of the political decision-making process but can become influential (Hogan and Doyle 2007).

One change explanation focuses on policy entrepreneurs accumulating enough power vis-à-vis those vested in and empowered by the extant institutional order. By overcoming the resistance of vested interests, which could block the reform process and act as veto players, policy entrepreneurs are able to steer the reform process (Capoccia and Kelemen 2007; Capoccia 2016b; Balleisen et al. 2017). Another explanation focuses on the presence of alternative ideas as an impetus for change. If policy entrepreneurs consolidate around one idea for an alternative institutional or policy design, a change in that direction is likely (Hogan and Doyle 2007; Birkland 2007; Baumgartner and Jones 1993; Blyth 2002). The case of Japan's nuclear safety governance reforms reveals that the convergence of agents of change, particularly political entrepreneurs, around one alternative idea can help overcome resistance by vested interests that previously exercised veto powers. Thus, this book argues that ideas and power can interact as change mechanisms during a critical juncture. While the exact relationship between the two mechanisms for change has remained understudied, political entrepreneurs clearly are at the nexus of policy ideas and political power (Béland, Carstensen, and Seabrooke 2016). Concretely, the findings here show that a united stance behind one idea, particularly in the case of collective political entrepreneurs who need to act in unison to exercise power, is what bridges power and ideas as change mechanisms. This helps to explain how a window of opportunity turns into a critical juncture, thus contributing to the ongoing debate about idea- and power-based explanations of institutional change.

Regulatory Agencies

This book argues that post-3.11 reforms rapidly shifted Japan's safety governance from aggressive promotion, including regulatory capture, to rigorous oversight. It challenges expectations that regulatory capture would reemerge after 3.11 due to powerful actors undermining new institutions. More broadly, it challenges the scholarship that sees capture as inevitable (Stigler 1971). Regulatory capture "refers to cases in which a regulated industry is able to control *decisions* made about that industry by regulators and/or *performances* by regulators related to the industry" (Mitnick 2011, 35). This regulatory capture definition highlights the decision-making process and the performance of regulatory agencies in areas where the stakes, that is, risks and benefits, are high. The more the industry can benefit from capturing regulation, the higher the incentive to attempt it. Building on Wilson (1980, 357), an early advocate of the idea that "there *is* a politics of regulation," the findings here show how independent safety governance can be institutionalized despite efforts to undermine it.

With the rise in independent regulatory agencies, the organizational characteristics of regulatory agencies have received renewed attention. Independent regulatory agencies are "institutionally and organizationally disaggregated from the ordinary bureaucracy and constitutionally separated from elected politicians" (Gilardi and Maggetti 2011, 201). Studies of their organizational characteristics increasingly pay attention to both formal and de facto independence (Hanretty and Koop 2013). Beginning with formal aspects, they denote the "primary dimension that political principles can control when delegating powers to regulatory authorities" (Gilardi and Maggetti 2011, 203), which pertains to the official rules regarding 1) the chairman and the management board, 2) the agency's formal relationship with elected politicians, 3) finances and internal organization, and 4) regulatory competencies. De facto independence "connotes the extent of regulators' effective autonomy as they manage their day-to-day regulatory actions" (Gilardi and Maggetti 2011, 203). Hence, it gauges how much autonomy an agency has in decision-making and whether politicians, regulatees, or other parts of the bureaucracy can influence agency decisions.

De facto independence in an agency's day-to-day interactions echoes Wilson's (1989) notion that staff exchanges between the regulator and the industry further the industry's ability to influence the regulatory agency by creating a revolving door between regulators and regulated interests. For Japan, this notion has to be extended to the ordinary bureaucracy, where staff rotation within and between ministries is a common practice that has previously inhibited expert knowledge allocation within the nuclear safety agency. Jordana, Fernández-i-Marín, and Bianculli (2018) add that an agency should be accountable to the public, while others go further and stress the need for public participation in a risk society to broaden the discussion about acceptable residual risks (Beck 1992).

Apparently, it is more difficult for nuclear safety agencies to attain independence, at least compared to agencies in other sectors. A large cross-sectoral study found that "nuclear safety agencies . . . are characterized by much less political independence, considerably less managerial autonomy and public accountability as well as slightly weaker regulatory capabilities than IRA in other sectors" (Jordana, Fernández-i-Marín, and Bianculli 2018, 535). In other words, nuclear safety agencies, on average, exhibit less formal and de facto independence. Regulatory capture, however, is not inevitable, even in the nuclear power domain.

Next to the abovementioned internal factors, there are certain external factors, such as audiences, public attention, and involvement in networks, that can help bolster agency independence. Agencies can "forge" their independence as "autonomy (or lack of it) is premised on its organizational reputation and the networks that support it" (Carpenter 2001, 4). To achieve autonomy, audiences are an important factor. Audiences refers to those individual or collective actors monitoring a regulatory organization, which include politicians, regulated interests, the media, and ordinary citizens (Carpenter 2010, 33). However, we cannot assume constant monitoring from ordinary citizens as public attention is fickle in the context of multiple policy issues vying for attention at any given time. If public interest is high and results in mobilization, it supports the existence of an independent agency (Wilson 1989). Some audiences are interested in nuclear safety by definition, such as international organizations and peers active in the same field. Involvement in regional or international networks strengthens an agency's domestic position (Eberlein and Newman 2008; Groenleer 2012; Bianculli, Jordana, and Juanatey 2017), because internationalization "provide[s] the agency a combination of technical knowledge, organizational practices, and legitimizing ideas" (Bianculli, Jordana, and Juanatey 2017, 1262). In short, reputation, an alert public, and international networks can function as a shield against external tampering with a regulatory agency's independence.

Drawing on scholarship on regulatory agencies, the findings on Japan's newly created nuclear safety agency highlight how internal and external factors are connected. A high degree of transparency is an important door opener for scrutiny by domestic and international audiences, which in turn helps shield agency independence. Furthermore, the findings stress the importance of determination by board members to actively create and retain agency independence based on their knowledge about the channels of influence provided by the institutional context in which the agency is embedded. Concretely, rules that forestalled the revolving door between the agency and the regulated interest as well as the ordinary bureaucracy were important factors in increasing the NRA's expertise and ability for independent decision-making.

The degree of discretion enjoyed by IRA defies the expectation in principal-agent theory that an agent with similar preferences to the principal is selected and that the principal will necessarily seek to exert control over the agent (Maggetti and Papadopoulos 2018). It is precisely this lack of political influence that can be in the interest of political decision makers in order to increase the credibility of a political commitment to new regulatory institutions (Gilardi 2008). Especially in the aftermath of a crisis, political actors seek to regain trust and legitimacy by limiting their own influence and that of their successors (Pierson 2000b), making the creation of an independent agency more likely. The desire to make a credible commitment can reinforce another effect, namely initial uncertainty about the outcome. At the time of decision-making, even the most strategic and well-informed actors may be motivated by short-term considerations, constrained in their choices, and at best hazily aware of later possible interactions with other institutional arrangements (Fioretos 2011; March and Olsen 1983; Pierson 2000b). Tensions between political intentions and eventual institutional outcome are illustrated well by the Japanese case, where nuclear safety reforms designed to regain domestic and international trust in order to enable the continued use of nuclear power have, in fact, eroded pronuclear actors' ability to implement nuclear policy and thereby ushered in a de facto nuclear phaseout.

Case Selection, Data, and Chapter Overview

This book adopts a problem-driven research strategy in the historical institutionalist tradition. It starts from Japan's recent regulatory governance reforms, with the puzzling observation that an independent nuclear regulator could not only emerge in Japan but also persist in the face of a strong pronuclear iron triangle pushing for a return to pre-3.11 business as usual. The approach in this book combines an analysis of a regulatory agency's organizational traits with an assessment of the political environment to explain the persistence of the NRA and its effect on nuclear politics. Given the country's long history of futile reform attempts and recent critical juncture in nuclear safety governance institutions, the Japanese case provides an empirically rich source for theorizing institutional change in the aftermath of focusing events.

As mentioned earlier, the NRA holds different functions, such as ensuring the safety of commercial nuclear power facilities, supervising radiation monitoring, safeguarding nuclear materials, and ensuring the safety of nuclear material. In the latter two areas, strongly intertwined with foreign and security policy, the NRA mainly translates provisions found in international treaties into domestic regulations. With regards to radiation monitoring, the NRA devises regulations while monitoring activities are conducted by other agencies and organizations. Nuclear safety, the one area the NRA fully controls, will be the focus of this analysis.

Studying critical junctures requires a historical perspective that includes the previous institutional pathway, a focusing event as a short-term cause, the critical juncture itself, an ensuing period during which institutions put down roots, and the newly emerging institutional pathway (Capoccia and Kelemen 2007, 351;

Collier and Munck 2017, 3). The present book accommodates this methodological requirement in its chapter structure. Following the "conjunctural causation"
turn in shock-induced change theory (Birkland 2007; Balleisen et al. 2017; Blatter and Haverland 2012), the book traces the origins, evolution, and effects of
nuclear safety regulation reforms.

Concretely, chapter 1 on Japan's "nuclear village" traces the history of nuclear
power policy and safety regulation until 2011. Chapter 2 covers 3.11 as a focusing event and the ensuing reforms. In order to overcome a lack of generalizabil-
ity stemming from a selection bias in favor of an outcome (e.g., critical juncture)
that qualitative case studies often exhibit (Blatter and Haverland 2012, 100–102),
this chapter increases the number of reform cases to two by comparing successful nuclear safety governance reforms with the aborted nuclear phaseout initiative. Chapters 4 through 6 cover different aspects of the post–critical juncture
phase. Chapter 3 takes an in-depth look at the new safety agency followed by an
assessment of newly introduced safety standards in chapter 4. The newly emerging pathway is covered in chapter 5 on the fissured "nuclear village."

Studying critical junctures means analyzing decision-making under conditions of uncertainty that were key in steering the institutional setting in one direction or another. The process tracing method is well suited to reconstruct
each step of the decision-making process, to identify influential decisions and
options available at the time (Capoccia and Kelemen 2007, 354–55). Considering available options as part of a counterfactual analysis involves considering
"what happened in the context of *what could have happened*." (Berlin 1974, 176,
italics added by Capoccia and Kelemen 2007). To understand the importance of
a single decision, the researchers need to assess "how close actors came to selecting an alternative option, and what likely consequences the choice of an alternative option would have had for the institutional outcome of interest"
(Capoccia 2016a, 92). Counterfactual analysis as a method to support a causal
explanation invokes other established regularities, theories, or principles found,
for example, in existing literature (Fearon 1991, 177). It may also involve a comparison with a hypothetically plausible alternative that feeds on the researcher's
background knowledge (Collier 2011, 825). Thus, a counterfactual analysis helps
to establish the respective historical weight of a certain decision in determining
the observed outcome by assessing whether another decision would have led to
a similar outcome or not.

Once it is apparent which decisions were historically important, process tracing also requires a researcher to identify the reasons why this decision was chosen rather than an alternative. Assuming boundedly rational actors, what is
deemed an appropriate decision differs from actor to actor and has to be uncov-
ered as part of the analysis. What actors deem appropriate depends on how they

understand the problem at hand and whom they blame for it. Problem definition and the inherent attribution of blame (Stone 1989) are a crucial element of the narratives different actors seek to construct in the aftermath of a focusing event (Birkland 2007; Blyth 2002).

Since the narrative that garners the most support will influence decision-making, understanding why a certain decision was taken is based on an analysis of the frames used by decision makers and which one prevailed. Framing analysis looks at three elements, namely diagnostic framing, prognostic framing, and motivational framing. The function of diagnostic framing is to identify the source of the problem at hand and focus responsibility or blame. Prognostic framing refers to articulating a proposal as to how the situation could be rectified, such as an appropriate decision. Diagnostic and prognostic framing are related since the nature of the identified problem tends to influence possible suitable solutions. Lastly, motivational framing serves to mobilize others by providing a rationale for engaging in collective action to address the situation (Benford and Snow 2000). Such framing analysis reveals why certain historical decisions were made by taking a closer look at the interpretations and motivations revealed by key actors.

Using process tracing, including counterfactual and framing analysis, to study a critical juncture requires a researcher to think more like a detective and less like a statistician. Process tracing necessitates thorough "soaking and poking" of a broad range of materials in order to gather information revealing key decisions, alternatives, perceptions, and motivations of actors, be they individual or collective decision makers (Blatter and Haverland 2012, 105–6). The analysis is informed by a variety of Japanese- and English-language materials, including official documents, biographical accounts, and interview data. Interviews with officials and stakeholders have served as background information to fill the blanks left by official documentation and to understand their motivation and interpretation regarding key decisions after 3.11. Expert interviews focused on active and former public officials and academics who were involved in nuclear regulation governance and its reforms. As is typical for such a methodology, the selection of interviewees mainly followed the snowball principle, where initial contacts provide further recommendations regarding who is experienced and willing to discuss regulatory and reform issues with outside researchers (Mosley 2013).

Nevertheless, the account this book offers is an interpretation that is inevitably selective (cf. Carr 1961). My main interest is the interaction between governance institutions constraining political actors and political agency as a driving force of institutional creation. The book therefore sets out with the establishment of Japan's nuclear policy domain, including regulatory governance institutions.

Chapter 1 tells the story of half a century of stable nuclear power development, from 1955 until 2011, and identifies the mechanisms that sustained path dependence. Stability rested on two prototypical pillars. One, the "nuclear village," consisted of Japan's mighty iron triangle of LDP, MITI, and big businesses. Its institutions structured the political conflict by limiting decision-making power to a few pronuclear actors and keeping the public out. The power of this nuclear iron triangle was buttressed by state intervention to support nuclear power and a selective information system that allowed it to influence public opinion. Much of the technical expertise and policy implementation power resided with electric utilities. Domestic nuclear incidents in 1974 and 1999 resulted in largely cosmetic nuclear safety governance reforms. They failed to separate nuclear power promotion and regulation, and they never touched electric utilities' decision-making power over accident countermeasures and thus safety costs. Second, widely accepted supporting ideas, such as the priority of economic growth and energy security, bolstered by the active promotion of a safety myth, increased public acceptance of nuclear power. Opposition remained limited to NIMBY protests against the siting of individual nuclear power plants. Hence, the nuclear village preserved the integration of nuclear power promotion and regulation, kept the public out of safety debates and in the dark about residual risks, and pushed nuclear power expansion as an affordable way to realize security, economic, and environmental policy aims. In the language of costs often used by historical institutionalist scholars, the nuclear village kept technical costs down at the expense of public safety and limited *social acceptance* costs by keeping the public and critics out while simultaneously becoming more invested in their own institutional arrangement over time.

Chapter 2 analyzes nuclear policy and regulatory governance reforms after the 2011 nuclear accident delivered a shock to Japan's energy system. Comparing the successful reform of nuclear safety regulation, a critical juncture, with the aborted initiative to phase out nuclear power, a so-called near miss, this chapter tests two hypotheses about alternative ideas and policy entrepreneurs' power as change mechanisms. First, the DPJ government, domestic experts, the public, international actors such as the IAEA, and even the LDP consolidated around the idea of a more independent nuclear safety administration and succeeded in pushing through reforms. Crucially, the DPJ government helped designated NRA board members and senior officials bring their vision of an independent agency to life before the LDP returned to government. Second, the nuclear village was able to defend its pronuclear policy in the absence of international pressure and because the DPJ government as a political entrepreneur was divided over whether to phase out nuclear power. The results show that convergence of collective political entrepreneurs around one idea is crucial because it can increase their power to shape reforms.

Chapter 3 assesses how the NRA put down roots as an independent regulatory agency. Many observers expected largely cosmetic reform outcomes similar to those after previous accidents. This time, however, the legal framework granted considerable authority to the agency, both over nuclear safety and its own operations. Using this managerial autonomy, the NRA board took the lead in institutionalizing independence, neutrality, and transparency early on. It put an end to previous practices, such as staff rotation between promoters and regulators and selective information release to influence public opinion. New procedures included building up in-house expertise, avoiding experts with close ties to the nuclear industry, opening meetings to the public, and making press conferences open to everyone. These findings provide insights into how a regulatory agency can avoid capture by building up independent in-house expertise, adopting transparency to enable domestic and international scrutiny, and attaining stronger competencies to enforce regulations.

Chapter 4 turns to the newly introduced safety standards. It focuses on identifying their strengths and weaknesses in comparison with previous standards and with IAEA recommendations. The new regulations constitute a significant improvement: for example, they include accident countermeasures and emergency response preparedness within a certain radius around a nuclear power plant. At the same time, the revisions leave room for improvement, as risk assessment methods utilized by the NRA lag behind international standards. The second part of the chapter turns to the implementation of safety standards. The NRA forced electric utilities to make far-reaching and costly improvements and punished those that sought ways to work around it. Not taking the strictest stance possible, however, NRA decisions exhibited a bias toward addressing the causes of the 3.11 nuclear accident and adopted a conventional rather than a precautionary risk regulation approach.

Chapter 5 returns to the nuclear policy domain and traces the simultaneous decline of the nuclear village and the expansion of antinuclear protests. After 2011, traditional promoters of nuclear energy remained in charge of energy policymaking, but the NRA undermined the mechanisms, the glue, that sustained their previous power in three ways. First, independent safety regulation substantially increased technical safety costs, particularly for electric utilities. Second, the NRA's transparency destroyed the information monopoly of the nuclear village, thus curtailing its influence over public opinion. Third, the NRA's emergency preparedness regulation expanded the metaphorical backyard of NIMBY protests, drawing in more protests. Thus, the shift from captured regulation to independent regulation has limited the ability of the nuclear village to control technical safety and social acceptance costs. After early pressure on the NRA came to nothing, the Abe government shifted to stressing its commitment to nuclear safety while providing lackluster support to the nuclear industry. As such, nuclear safety reforms after 3.11 undermined the feasibility of nuclear power amid weakened political and administrative support. Rising technical safety costs and further social acceptance costs are disproportionately borne by the nuclear industry, especially electric utilities, which significantly changed their business environment for the worse. In essence, safety reforms caused fissures inside the nuclear village and undermined the power of the pronuclear coalition to implement nuclear power policy.

The concluding chapter returns to the puzzle of an independent nuclear regulation agency with the ability to withstand pressure from the pronuclear iron triangle. The answer to the puzzle lies in the initial steps taken immediately after the accident, in line with historical institutionalism's focus on temporality. The results reveal that safety governance reforms can become political game changers if they undermine the power structure that sustained the previous path. Thus, theories about crisis-induced change should be more specific about which institutions matter in a given policy domain with regards to both policymaking and implementation. Regarding change mechanisms, the results demonstrate the importance of agents of change agreeing on one reform idea. As political entrepreneurs consolidated around the idea of independent safety regulation, power constellations changed in favor of those seeking change. Thus, a united stance among collective (political) entrepreneurs behind one idea is what bridges power and ideas as change mechanisms. Independent regulation and transparency are of particular practical relevance, not least because of their impact on technical safety costs and social acceptance costs. For the governance of high-risk technologies, this chapter distills the findings to a few central lessons. As the well-known proverb says, "knowledge is power." Avoiding regulatory capture requires state expertise and regulatory competencies vis-à-vis developers and promoters of a technology, as well as informational transparency as a door opener for a societal debate about acceptable risks and scrutiny of a regulatory agency's work.

THE "NUCLEAR VILLAGE"

The accident at the Fukushima Daiichi nuclear power plant in March 2011 (simply referred to as "3.11" in Japan) naturally drew attention to Japan's nuclear power program. Japan's history of promoting the civilian use of nuclear power to generate electricity dates to the post–World War II period as part of the effort of resurrecting the country from the ruins of war. By 2011, Japan had fifty-four commercial nuclear reactors, making it the country with the third largest number of nuclear reactors, following the United States and France. The expectation inside and outside the country had been that nuclear power was in safe hands in high-tech Japan. The 3.11 events led people to ask how such a disaster was possible.

Questions about the accident's cause triggered inquiries into the wheelings and dealings of powerful actors that promoted nuclear power. In the late 1990s, Iida Tetsunari[1] coined the term "nuclear village" to describe the "syndicate" of actors pushing Japan's nuclear power program (Asahi Shimbun 2012a, 13). This term has since been used, mainly by critics, to describe "institutional and individual pro-nuclear advocates in the utilities, the nuclear industry, the bureaucracy, the Diet (Japan's parliament), business federations, the media, and academia. This is a village without boundaries or residence cards, an imagined collective bound by solidarity over promoting nuclear energy" (Kingston 2014, 108).

Others disagree with the picture it evokes of everybody in the "village" acting in concert (Shiroyama 2015; Hymans 2011), criticize its increasingly condescending use (Scalise 2013), or avoid it altogether (Kikkawa 2012). This book uses the term because it found its way into political debates in post-3.11 Japan. As the nuclear village was made up of collective actors, it was only natural that not every individual member was always in agreement. Hence, this book stresses fuzzy boundaries and conflicts among pronuclear actors in Japan, as in a real village with family feuds.

A prominent feature of the nuclear village was regulatory capture, that is, state regulators devising rules in the interest of the industry rather than the public interest (Kingston 2014, 110; Ferguson and Jansson 2013; Vivoda and Graetz 2014, 10). In the same vein, the Nuclear Accident Independent Investigation Committee, established by the Diet, concluded that "the TEPCO Fukushima Nuclear Power Plant accident was the result of collusion between the government, the regulators and TEPCO, and the lack of governance by said parties" (NAIIC 2012, 16). Japan's nuclear policy history is sometimes divided into stages with the restructuring of the nuclear power administration in the mid- to late 1970s and in the early 2000s marking turning points (Kikkawa 2012; Shiroyama 2012; Kishimoto 2017). This chapter shows that these reforms only changed the makeup of the nuclear safety administration but left intact the safety governance flaws that ailed the nuclear village.

Understanding such path dependence in the nuclear power domain requires a closer look at the institutions that structured it. Institutions, the humanly devised "rules of the game" that structure interactions of different organizations (North 1990), encompass "the formal and informal procedures, routines, norms and conventions embedded in the organizational structure of the polity or political economy" (Hall and Taylor 1996, 938). Institutions and organizations are clearly linked in political practice but need to be distinguished for analytical clarity. Hence, this chapter studies collective actors (organizations) in the nuclear power domain and the formal and informal rules (institutions) that structure their interaction to explain the emergence and persistence of safety governance institutions that ultimately led to the March 2011 nuclear disaster.

Institutions, and the organizations they are embedded in, determine actors' respective power. According to Lukes (1974), power in politics has three faces. First, agenda-setting power is the ability to put an issue on the political agenda or keep it out of the spotlight. Second, decision-making power refers to the ability to decide decision-making for oneself or to prevent a decision from being taken at all. Third, depoliticization power refers to the ability to convince others to accept one's own interest as their own and to refrain from attempting to put a topic on the political agenda in the first place. Institutions determine who is able to set the agenda, influence decisions, and play a role in politics.

The power structure present at the time of institutional formation is codified in stable institutional arrangements (Fioretos 2011). The distributional nature of institutions empowers some actors while marginalizing others, creating winners and losers of a particular institutional order. Empowered actors will have an incentive to sustain the arrangement and their position while those who are marginalized have an incentive to seek participation (Thelen 1999). Within a given policy domain, for example nuclear politics, the distributional nature of institutions translates into a rather stable actor constellation. Those in decision-making positions, the winners of an institutional order, constitute a "policy monopoly" (Baumgartner and Jones 1993, 7) or "dominant coalition." Usually, a policy domain consists of at least one dominant and one minor coalition (Sabatier and Weible 2007, 196). Over time, actors adapt their behavior to a given institutional arrangement in ways that reflect it and reinforce the underlying logic of the system. There are increasing returns, because the costs of switching to a previously plausible alternative increase with each step down a chosen pathway, decreasing the likelihood of completely changing paths (Pierson 2000a). Due to vested interests and increasing returns, the incentive for the winners of an institutional order to defend it against challenges only grows with time.

An understanding of how post-3.11 nuclear safety administration reforms finally broke with path-dependent safety governance practices and the domination of nuclear politics by the nuclear village requires a clear picture of that path. This chapter sketches out the institutional arrangement that supported it. It begins in the postwar period when the administration for Japan's nuclear power program was established. A closer look at Japan's nuclear energy policy and safety history—more than fifty years of remarkable policy stability, even across two instances of organizational restructuring—highlights how the nuclear iron triangle sustained the institutional arrangement it collectively profited from, while cost-related feuds emerged that would become the Achilles' heel of the pronuclear coalition after 3.11.

Early Industry Opposition to State Control

World War II and its immediate aftermath constituted a period of institutional formation for Japan. Decisions were made that shaped Japan's democratic institutions. As many things were in flux, there were struggles over decision-making powers within policy domains. The main objective for decision makers was to rebuild the country. In the mid-1950s, a sustained period of rapid economic growth began, with real GDP growth rates averaging around 9 percent annually until the 1973 oil shock. This growth presented a challenge for the energy sector as it created a rapidly increasing demand for energy. Nuclear power was viewed as an exciting future technology that promised an endless electricity supply. This was true not only in Japan but also in other countries, such as the

United States, Germany, and France. In the early 1950s, US President Eisenhower began promoting the peaceful use of nuclear materials in commercial nuclear power plants. The Atoms for Peace program supplied information and equipment to US allies around the world, including Japan.

Japan's political decision makers sought to shape the electricity industry in their image, reigniting a conflict over the state's role in the electricity industry that dated back to wartime. To support war efforts, militarists and bureaucrats sought to nationalize electric utilities, which they effectively achieved by 1941, despite adamant industry opposition to bureaucratic interference (Johnson 1978, 132–34). Under US occupation after World War II, a system of nine utilities[2] was formed in 1951. To represent their interests, utilities formed the Federation of Electric Power Companies (*Denki-jigyō-rengōkai*, short *Denjiren*). Denjiren went on to become the mouthpiece of the regional electricity monopolies, with Japan's largest utility, TEPCO, as a powerful voice within it. With economic growth as a political priority, the government attempted to reshape the nine regional monopolies into one public corporation to meet the interests of big businesses seeking subsidized power. MITI supported the government move, because it was also in favor of a centralized, coordinated, and subsidized power industry. But this time electric utilities successfully opposed state control. This established an important precedent and, henceforth, MITI refrained from challenging the utilities in such a manner (Samuels 1987, 159–62). Importantly, the system of regional electricity monopolies persisted, and they went on to become influential actors in the nuclear village.

In the beginning, the LDP attempted to create hierarchical administrative organizations for nuclear power. The 1955 Atomic Energy Basic Law (*Genshiryoku-kihon-hō*) created the Atomic Energy Commission (AEC, *Genshiryoku-i'inkai*) and the Science and Technology Agency (STA, *Kagaku-gijutsu-chō*), both located in the Prime Minister's Office. The AEC chairman would also serve as STA director, a ministerial position, giving the double agency head considerable weight in nuclear policy. The AEC was in charge of formulating policy for a peaceful use of nuclear power. STA, in turn, acted as a secretariat to the AEC, assisted the AEC in policy formulation, and translated policies into concrete research and development directives. For research and development, STA oversaw the Japan Atomic Energy Research Institute (JAERI, *Nihon-genshiryoku-kenkyūjō*), a semigovernmental organization with the task to further research on nuclear power. Since both the AEC and STA reported directly to the prime minister, this system was designed to facilitate nuclear power development under political leadership.

The nuclear industry however, favored a less hierarchical approach and found a powerful ally in the first head of the AEC and STA, Shōriki Matsutarō. He brought the regional monopolies and the heavy industries together under the banner of the Japan Atomic Industrial Forum (JAIF, *Nihon-genshiryoku-sangyō-kyōkai*). Its purpose was to establish a common position between the two industry groups and to promote the civilian use of nuclear power. This move was an important step to make the industry a full partner and to integrate it into the policymaking process. In JAERI, too, Shōriki made the industry a partner. In exchange for assuming part of the budget, it was able to lead research and development (Hymans 2011, 165–66). Hence, the government relied on industry expertise to develop indigenous nuclear technology.

Shōriki was also president of the Yomiuri newspaper company, and his stewardship turned the *Yomiuri Shimbun*[3] and TVAsahi into pronuclear news outlets. The aim was to counter antinuclear sentiments after the devastating atomic bombings of Hiroshima and Nagasaki. Shōriki's involvement in the promotion of nuclear power on many fronts led to the moniker "father of nuclear power" (*Genshiryoku no chichi*) (Arima 2008, 12–13). Shōriki's legacy was a nuclear industry as an equal partner in policymaking and research, as well as a pronuclear mass media presence.

Somewhat surprisingly, MITI played only a minor role in early nuclear power development, even though it was the ministry at the heart of postwar state-led economic planning to catch up with the West, most well known under the concept of the "developmental state." As such, MITI was later credited with Japan's economic rise (Johnson 1982). Due to industry opposition to bureaucratic interference and political top-down leadership, however, the electrical industry developed differently than those directly fostered by the government (Johnson 1978). With regard to nuclear power, competencies MITI would have liked to have were handed to STA instead, leaving it on the losing side of early political battles. MITI nevertheless supported nuclear power development and the nuclear industry (Hymans 2011, 171). In 1969, MITI also created the Japan Atomic Energy Relations Organization (JAERO) as a lobbying organization for nuclear power. MITI was not a central player in the nuclear domain yet, but it promoted Japanese nuclear power development as a matter of national pride with high symbolic value as a technology initially developed by the industrialized countries Japan was trying to catch up with.

With all these factors in view, the Japanese government began to lay the groundwork for developing "nuclear energy for a bright future."[4] These early struggles over the state's role in the nuclear energy program shaped state-industry relations in the nuclear policy domain at a time when institutions were malleable. Importantly, the industry successfully opposed politicians' attempts to create one public corporation and to establish top-down state leadership. Instead, the nuclear industry became a partner with a say in nuclear policy and related research. The first reactors to take up commercial operation were at the Japan Atomic Power

Company's Tōkai nuclear power plant (NPP) in 1966 and Tsuruga NPP in 1970. They were followed by Kansai Electric Power Company (KEPCO) with the Mihama NPP in 1970 and Tokyo Electric Power Company (TEPCO) with the Fukushima Daiichi NPP in 1971. The common goal was to quickly build a nuclear infrastructure and to put in place an entire nuclear fuel cycle. This early period crucially shaped the path nuclear power development would follow for half a century: a nuclear industry as a partner rather than a regulated party and state support without much state control.

The Nuclear Iron Triangle

The 1973 oil shock shook Japan's energy policy and gave nuclear power an additional push. Global oil prices quadrupled, which dealt a blow to Japan's postwar energy policy. Imported oil, which was cheaply available on the world market, had become the most important fuel supporting postwar economic growth. With more than 50 percent of the power generating capacity being oil fired, increasing oil prices exposed how vulnerable the Japanese economy was to instabilities in global oil supplies. The experience was comparable to what could be called an energy trauma. Decision makers' reaction was to diversify energy sources, including the promotion of nuclear power as a quasidomestic source of energy[5] in order to reduce Japan's import dependence and to provide stable and affordable electricity for Japan's economic growth.

One year after the first oil crisis, the *Mutsu* nuclear-powered ship accident cast doubt on the safety of nuclear power. During the ship's high-profile launch in early 1974, there was a radiation leak aboard. Reports of this leak led to a safety debate. MITI successfully attacked STA's competence in the aftermath of the *Mutsu* incident (Hymans 2011, 171) and assumed responsibilities related to the implementation of nuclear policy. This left STA weakened but still in charge of research and development, including oversight of the first fast breeder reactor, Monju. In 1974, MITI established the Agency of Natural Resources and Energy (ANRE, *Shigen-enerugī-chō*) to help formulate Japan's energy policy, including nuclear energy. Since then, the AEC and ANRE have shared the task of nuclear energy policy formulation. As a former AEC official put it, the AEC went from being a consensus broker between the nuclear industry and the government on long-term plans to formulating guidelines for MITI (interview with former AEC official, 2015). To implement nuclear policy, ANRE was put in charge of granting licenses for construction, testing, and operation of nuclear reactors. With MITI at the center, the licensing process was streamlined, and the number of licenses required decreased from 160 to only 66 (Cohen, McCubbins, and Rosenbluth 1995, 183). A safety bureau within ANRE conducted necessary safety assessments. Administrative restructuring following the 1973 oil shock moved MITI into the center of nuclear policymaking, licensing, and safety regulation.

Partially Voluntary Safety Regulation

In response to the *Mutsu* nuclear accident, the AEC formulated the "Regulatory Guide for Reviewing Seismic Design of Nuclear Power Reactor Facilities" in 1978 with new safety requirements that nuclear reactors had to meet in order to gain an operating license. These safety requirements regulated a reactor's earthquake resistance, something plant manufacturers had to take into account when designing and building reactors. Other natural disasters, such as volcanic eruptions or tsunamis, and accident countermeasures were not subject to regulation. There also was no regulation about in-use standards or refurbishing nuclear reactors in line with new scientific findings about necessary safety measures, known as backfitting. Limited safety standards left it to the discretion of each electric utility to backfit reactors and to implement disaster countermeasures. The AEC's 1978 standards imposed safety regulation on plant manufacturers much more than on electric utilities as operators. This allowed utilities to operate a nuclear reactor for sixty years once an operation license was granted and to determine safety costs to a large extent.

Also, in response to the *Mutsu* nuclear accident, the government set up an advisory committee to reexamine the structure of the nuclear safety administration. Following the committee's recommendation to separate nuclear safety functions and nuclear policymaking functions, the government transferred nuclear safety functions out of the AEC and into a separate commission within the Prime Minister's Office (Kishimoto 2017, 316–17). This new Nuclear Safety Commission (NSC, *Genshiryoku-anzen-i'inkai*) remained weak, however, and double-checked investigations by ANRE's safety bureau rather than conducting its own investigations. It constituted a toothless body that made recommendations and nonbinding guidelines (Hymans 2011, 171; Ferguson and Jansson 2013). The practice of partial industry self-regulation remained untouched by the establishment of the NSC in 1978. Despite the intention to strengthen nuclear safety by separating the promotion and regulation of nuclear power, creating the NSC failed to establish independent safety regulation.

As a result of ubiquitous staff rotation within ministries as well as between ministries and agencies at the Cabinet Office, both ANRE's safety bureau and the NSC were partly staffed with pronuclear public officials. Staff rotation allowed MITI to exercise influence over the NSC and limited a buildup of expertise within NISA. MITI officials transferred in and out of the regulating agency about every two years in line with regular staff rotation, and only a few had a background in nuclear engineering. Furthermore, the practice of *amakudari* (literally "to descend from heaven") meant that many MITI officials, who rotated through the safety agency NISA as part of their career path, would later find retirement jobs in the nuclear industry. Thus, there was little incentive for them to enforce stricter and, thus, more costly safety requirements on the nuclear industry. Staff rotation prevented a buildup of necessary technical expertise and did not encourage public officials to take responsibility for regulation, or a lack thereof. Furthermore, staff rotation gave MITI room to influence the NSC and provided little incentive for regulators to challenge the status quo of partial industry self-regulation. With administrative restructuring in the 1970s, MITI gained influence over the NSC, which effectively shifted limited regulatory competencies from the Prime Minister's Office to MITI and failed to substantially improve the competencies of nuclear safety regulators vis-à-vis the industry.

State Support

Nuclear power expansion was not entirely uncontroversial. Already in the late 1960s, local protests against the siting of nuclear facilities emerged. Since the political system offered few entry points in the procedures for the construction and operation of a nuclear power plant, which was basically controlled by the utility in question, MITI, the AEC, the NSC, and the prime minister, opposition mainly took the form of not-in-my-backyard activism (NIMBY) (Cohen, McCubbins, and Rosenbluth 1995, 197). NIMBY activism increased after the 1974 *Mutsu* accident. In some cases, opposition groups garnered the support of local fishing and agriculture cooperatives and were able to prevent the siting of a nuclear power plant (Aldrich 2010).

To achieve the goal of diversifying Japan's energy supply and to push nuclear power expansion, the LDP passed the Three Laws for Electricity Generation (*Dengen-sanpō*) in 1974. Administered by MITI, the laws marked the beginning of the government systematically subsidizing the siting of nuclear power plants by compensating local opponents, such as fishermen, farmers, and local governments (Cohen, McCubbins, and Rosenbluth 1995, 186). Along with financial compensation for opposition groups, the scheme also included education on the benefits of nuclear power. For example, MITI's JAERO offered free classes and seminars to schools and communities open to hosting a nuclear power plant (Aldrich 2014, 85–86). Thus, in addition to ANRE streamlining the licensing process for electric utilities, the *dengen-sanpō* offered education on the benefits of nuclear power and side payments to ease local opposition to the construction of nuclear power plants. Kikkawa (2012, 57) dubbed this system of government intervention *kokusakumi-*

nei, which translates into "private implementation of public policy." The elaborate system of political and administrative support for nuclear power expansion and waste management was designed to help the nuclear industry carry out the government's energy policy plans.

Reciprocal Consent

In the 1970s, Japan's iron triangle—comprising the elite bureaucracy, most notably MITI, the LDP, and business interest groups—was at the heart of Japan's pronuclear policy domain. In the long-standing debate about who governs Japan—the bureaucracy, politicians, or the private sector—the famous characterization of Japan as a "developmental state," in which MITI (now Ministry of Economy, Trade, and Industry, METI) effectively orchestrated Japan's postwar economic recovery using "administrative guidance" (Johnson 1982, 1995) is possibly the most well-known example of the argument that the Japanese elite state bureaucracy dominates policy decision-making (Pempel 1982; Vogel 1999). Regarding potential bureaucratic dominance in the nuclear power domain, Johnson stressed that electric utilities were strongly opposed to government controls and "developed differently from industries that have been directly fostered by the government" (Johnson 1978, 132).

The counterview in the debate about who governs was the model of politicians as "principals" and the bureaucracy as their "agents," which puts politicians in the driving seat of Japanese politics (Ramseyer and Rosenbluth 1993). In case of nuclear power, the Prime Minister's Office, with STA and the AEC, was in charge of a framework for civilian nuclear power development. In the next step, "once government approval, most of the time LDP approval, had been given, the implementation was left up to MITI in the form of granting licenses" (Cohen, McCubbins, and Rosenbluth 1995, 195). MITI streamlined the licensing process for the nuclear industry, and the safety bureau within ANRE refrained from developing new safety standards, at least until a string of accidents beginning in the late 1980s. The relationship was one of "reciprocal consent," where the bureaucracy "achieves jurisdiction and is denied control in an iterative process with private interests within an environment of unusually stable elites and extraordinarily durable institutions. The Japanese bureaucracy does not dominate, it negotiates" (Samuels 1987, 260). Under these conditions, electric utilities held a position of power within the pronuclear coalition.

The industry, in turn, was a generous supporter of the LDP through campaign financing. Over time, over 70 percent of individual donations to the LDP came from current or former senior officials at TEPCO and other utilities (Lukner and Sakaki 2011, 52). Business interests were integrated into the policymaking process through business federations. The manufacturing business association, Keidanren, and the electric utilities association, Denjiren, played important roles as mediators between the nuclear industry and MITI. Furthermore, JAIF functioned as a mechanism to forge a consensus between electric utilities and plant manufacturers. Bringing businesses and bureaucracy even closer together was the already mentioned *amakudari* practice of bureaucrats taking up well-paid retirement posts in the industry. Once sanctioned by the LDP, implementing nuclear policy was effectively up to MITI and the nuclear industry.

A crucial element was massive state support for nuclear power development without state control. This was a manifestation of state dependence on the nuclear industry to develop indigenous nuclear technology and to implement energy policy goals. In return for policy implementation, the MITI-administered *dengen-sanpō* scheme served to ease opposition and shape public opinion in favor of nuclear power through the system of subsidies and education on the benefits of nuclear power. While MITI held jurisdiction over nuclear energy policy and supported the siting process, it refrained from interfering with the nuclear industry, particularly electric utilities. This fit the third perspective on the iron triangle, which stresses that business interests win, at least during times of "quiet politics," when the public pays little to an issue and there is low voting salience (Culpepper 2010).

During the 1970s and 1980s, business conditions for nuclear power development were ideal. Electric utilities enjoyed state support and uncontested regional monopolies that allowed them to invest in new nuclear reactors. Japanese manufacturing heavyweights, such as Hitachi, Mitsubishi, and Toshiba, supplied nuclear reactors to the electric utilities. Limited safety regulation and state support made nuclear power development a profitable business for the utilities. The AEC's attempts to introduce more stringent regulation after the *Mutsu* ship accident imposed certain earthquake-related safety requirements on manufacturers but left utilities' freedom to self-regulate intact. About nuclear power expansion under these conditions, Cohen, McCubbins, and Rosenbluth (1995, 193) wrote that "because of the form of utility regulation, such expansion is profitable and virtually risk-free."

The pronuclear coalition succeeded in developing one of the most advanced civilian nuclear energy programs in the world. It aimed at a complete nuclear fuel cycle with plans for fast breeder reactors, mixed-oxide fuel use, and nuclear fuel recycling. The majority of Japan's nuclear reactors were built between 1974 and 1994. These included big projects, such as TEPCO's Fukushima Daiichi NPP, where five reactors were added; the Fukushima Daini NPP with four reactors; and the Kashiwazaki-Kariwa NPP, where five out of eventually seven reactors went into operation. The second biggest utility, KEPCO, completed four reactors at the Takahma NPP, another four at Ōi NPP and added one more to the existing two at Mihama NPP. Kyushu Electric joined the club of utilities operating nuclear reactors with four at the Genkai NPP and two at the Sendai NPP, as did Chubu Electric with four reactors at the Hamaoka NPP and Tōhoku Electric with two reactors at the Onagawa NPP. During the heyday of nuclear power development, the two biggest utilities, TEPCO and KEPCO, substantially expanded their nuclear reactor portfolios and some of the smaller utilities joined in the business.

Opposition

Beginning in the late 1970s, numerous accidents began to highlight that nuclear power was not risk free for people living around it. Outside of Japan, high-profile events occurred in 1979 at Three Mile Island in the United States and in 1986 at Chernobyl in what was then the Soviet Union. According to a long-running survey conducted by the *Asahi Shimbun*,[6] public support for nuclear power dipped after Three Mile Island in March 1979 but quickly recovered following the second oil shock later the same year. After the Chernobyl disaster in 1986, public opinion reversed for the first time. As public support for nuclear power began to recover in the mid-1990s, a string of smaller nuclear incidents happened in Japan. There were the 1995 natrium leak at the prototype fast breeder reactor Monju, the 1997 heavy water leak at the advanced thermal reactor Fugen, and the 1999 criticality event at the Tōkaimura JCO reprocessing plant. They not only fueled public doubts about the safety of nuclear power (Shibata and Tomokiyo 2014, 9) but also caused a reaction in civil society.

Japan's civil society was comparatively weak since the "laws governing the nonprofit sector made it extremely difficult for groups to obtain non-profit status" (Schreurs 2003, 244). While small groups, such as neighborhood associations, enjoyed state support, it was "hard for autonomous groups to become large and hard for large groups to be autonomous" (Pekkanen 2003, 133). Consequently, antinuclear protests were limited to NIMBY protests against the development of nuclear power plants. Few in numbers at first, a "new wave of anti-nuclear protests" emerged after Chernobyl in 1986 (Honda 2005; Hasegawa 2004). These NIMBY protests were network based and characterized by their rejection of bureaucratic organizational structures (Hasegawa 2004). In response to—sometimes successful—protests, the Japanese state displayed a "Machiavellian" stance by seeking to shape social preferences to serve its policy choices and by avoiding areas with stronger potential for resistance in the development process (Aldrich 2010).

One route for local opposition was to file lawsuits through the courts. The principal-agent relationship that put politicians in the driver's seat included judicial manipulation (Ramseyer and Rosenbluth 1993), which refers to the government using the appointment of judges as a way to punish those who ruled against the LDP's stance, at least on high-profile partisan disputes, for example the status of Japan's military forces (Ramseyer 2019). While courts presented an avenue for citizens pushing for stronger pollution regulation (Upham 1987), none of the lawsuits seeking to halt nuclear power development before 2011 were successful. According to a list of court cases and respective rulings (CNIC, pers. comm.), a total of fifteen lawsuits were filed against nuclear power plants between 1973 and 2010.[7] These lawsuits targeted the operating licenses of nuclear power plants or sought to impose a ban on operations. The outcomes were usually the same. As a rule, all lawsuits with a ruling before 2011[8] were eventually decided in favor of the state or the respective electric utility (interview with H. Ban, 2018), which appears to be in line with Ramseyer's (2019) findings that the judiciary will support the LDP's policy line on contentious issues. Nevertheless, local opposition against the siting of new plants had an impact on nuclear power development by causing delays in power plant construction and cost overruns for the electric utilities.

The Safety Myth

Public attention shifts and intensifying local opposition made political support and a positive image for nuclear power ever more important for the industry and its governmental supporters. Documents classified at the time, but declassified after 3.11, revealed that the Japanese government responded to the Chernobyl nuclear accident with a plan to reaffirm nuclear power for Japan. Shortly afterwards, it hosted a G7 meeting in Tokyo, at which it made sure that safety concerns were removed from the draft statement, which instead mentioned nuclear power as "an energy source that will be ever more widely used in the future." An official from the Ministry of Foreign Affairs involved in the meeting told the *Japan Times* that "There was no awareness in the government or the nuclear industry that Japan's nuclear plants might be dangerous too, or that we could learn a lesson from [Chernobyl]." Following the G7 meeting, ANRE sent out a note to electric utilities and local governments about the national government's intention to "continue to promote [nuclear power] with a safety-first mindset" (Japan Times 2017) instead of actually reexamining the safety of Japan's nuclear power plants.

The government's reaction was emblematic of what is known as the "safety myth" regarding the absolute safety of nuclear power plants made and operated in Japan. This myth was later identified as one cause of the Fukushima nuclear accident by the Parliament-sponsored Nuclear Accident Investigation Committee (NAIIC). In an NAIIC hearing, nuclear safety expert Madarame Haruki, a long-time member of the NSC, testified that the NSC had a culture of complacency. This included disregarding studies about possible large earthquakes off the coast of Fukushima Prefecture. He stated, "Though global safety standards kept on improving, we wasted our time coming up with excuses for why Japan didn't need to bother meeting them" (New York Times 2012). The blind belief of many in the "nuclear triangle"—operators, regulators, and politicians—in the country's technical mastery resulted in a failure to properly assess risks.

Testifying in a court case against Hamaoka NPP in the mid-2000s, Madarame had brushed away plaintiffs' concerns about a power cut putting a stop to the cooling system, which could result in a meltdown. In a nutshell, this is what happened at the Fukushima Daiichi NPP in March 2011. Back then, he said that worrying about all the things that could go wrong would "make it impossible to ever build anything" (New York Times 2011) After numerous hearings, the NAIIC concluded: "Because the regulators and operators have consistently and loudly maintained that 'the safety of nuclear power is guaranteed,' they had a mutual interest in averting the risk of existing reactors being shut down due to safety issues, or of lawsuits filed by anti-nuclear activists" (NAIIC 2012, 43).

As accidents sparked public safety concerns and protests, the promotion of the safety myth and lack of safety debate became crucial to garner public support. Adopting an attitude of "That which must not, cannot be"[9] turned the system established to promote nuclear power's civilian use for "a brighter future" into a tool to garner public support by downplaying the risks involved.

Some nuclear experts sought to make improvements to accident countermeasures and to reduce risks revealed by accidents and newer scientific studies. In the early 1990s, ANRE warned the utilities about the risk of cooling systems shutting down as the result of a power station blackout after a catastrophic event. It urged them to develop backup systems to keep cooling systems running even in the case of such an event. However, the utilities successfully opposed this, arguing that the existing backup systems in place were sufficient and that nuclear facilities were safe (Kingston 2014, 111). With utilities opposing accident countermeasures and ANRE lacking the competencies to enforce them, the safety myth prevailed.

In addition to the safety myth, expertise-related features of the nuclear safety administration hampered safety improvements. People with a critical view of nuclear safety have always been part of the nuclear safety administration, also inside METI, but it was their unquestioning colleagues who went up the career ladder into decision-making positions (interview with Environmental Policy Expert, 2014). There is no systematic research on the topic, but stories abound about scientists at universities who spoke out about risks and lost access to important means of research, such as research funding from utilities, manufacturers, and associated foundations. As a result, they also failed to proceed up the career ladder into positions where they might be consulted by the government. Those critical scientists who existed were either silenced in such a way or they left the nuclear village (interview with Nuclear Politics Expert, 2015). One scientist who left was Iida Tetsunari, the outspoken critic of nuclear proponents who coined the term nuclear village in the late 1990s. One reason he cited was the tendency to silence critical voices inside member organizations in their pursuit of developing nuclear power and to protect themselves from outside scrutiny (Asahi Shimbun 2012a, 13). Rather than giving critical experts a voice and regulators the necessary competencies, those in positions of power within the nuclear village disseminated the absolute safety myth to garner public support.

An Intra-Industry Feud

Faltering public acceptance was not the only challenge the nuclear industry faced. In line with an overall trend to liberalize sectors of Japan's economy, a stepwise deregulation of the electric sector began in the mid-1990s with the first step pertaining to large consumers. Even though the electricity market effectively remained monopolized (DeWit 2014, 122), this deregulation introduced a new logic into the system. Electricity market liberalization and promoting nuclear energy are contradictory in two ways. First, liberalization equals a retreat of the state, which stands in contrast to nuclear power development supported by state intervention. Second, liberalization requires a utility to think like a strategic actor in a competitive environment rather than a member of a cooperative network of actors. With state support for nuclear development on the one hand and partial electricity market liberalization on the other, electric utilities ended up "being caught in the middle" (Kikkawa 2012, 113–14). While utilities had to become more cost conscious, the same cannot be said for plant manufacturers. In effect, partial market liberalization drove a wedge between the two nuclear industry groups (Hymans 2011, 178). With manufacturers and operators not profiting equally anymore, fissures within the nuclear industry emerged.

The 1990s saw an increase in the costs of developing nuclear power. When the costs for the fast breeder reactor Monju exploded, tensions emerged between the industry and the administration. Eventually, the nuclear industry abandoned the Monju project, and the administration took over the costs associated with establishing a full nuclear fuel cycle. The nuclear industry was unable to prevent the partial market liberalization at this point, but it left the costly Monju project, and utilities successfully opposed ANRE's calls for additional costly accident countermeasures. At the same time, partial market liberalization began to erode the unified position of the nuclear industry. Similarly, increasing local opposition put more financial pressure on electric utilities than on plant manufacturers, because utilities had to shoulder the cost overruns of construction delays. The strategy adopted was to work around local protests against the siting of new nuclear power plants by expanding existing facilities. After sixteen new sites were developed between 1966 and 1994, the creation of new sites came to a halt in the second half of the 1990s. Instead, twenty-two reactors were added to existing sites. Cost issues revealed themselves as a potential Achilles' heel of the nuclear triangle by causing frictions and slowing down nuclear power development.

A Nuclear Renaissance?

The early 2000s saw another restructuring of the nuclear policy domain in the context of the state administration reforms, known as the Hashimoto reforms, which streamlined the Japanese administration by reducing the number of ministries and strengthened the Cabinet Office with new leadership instruments. Regarding nuclear safety administration, STA, the agency located in the Prime Minister's Office, was abolished. It had lost credibility due to its handling of the 1995 Monju sodium leak, which made it politically vulnerable (Hymans 2011, 175–76). Some of its tasks, such as oversight of the fast breeder reactor Monju, were handed over to the Ministry of Economy, Trade, and Industry (METI, formerly MITI). Removing STA left the AEC weakened as its head no longer held a function as a state minister. Also, the AEC staff was reduced to thirty, about half its previous size. In contrast to the overall strengthening of the prime minister's role, reorganizing the administration actually weakened the Cabinet Office's role in the nuclear domain.

Since the 1970s, both the AEC and ANRE had formulated nuclear policy, but the 2003 Basic Energy Law put METI's ANRE solely in charge of overall energy policy formulation and tilted the power balance in favor of METI. The AEC went from making policy plans, to formulating guidelines in the 1970s, to merely setting the framework in the 2000s. A former member illustrated AEC's diminishing decision-making power with a boat analogy. The AEC used to be a powerful tugboat for the nuclear domain. Over time, the boat it was trying to pull turned into a huge ship. When the Hashimoto reforms weakened the AEC, it resulted in a situation where the ship it was trying to pull, METI and the nuclear industry, would stay its course, regardless of the tugboat's directions (interview with Former AEC Official, 2015). Reducing the AEC's political weight and moving more STA functions to METI further shifted nuclear energy policy decision-making power from the Cabinet Office to METI.

Following the Hashimoto reforms, research and public relations organizations were reorganized as well. JAERI and the Japan Nuclear Cycle Development Institute were combined in 2005 to form the JAEA, which was responsible for research related to commercial reactors and the establishment of a full nuclear fuel cycle. The new Ministry of Education, Culture, Sports, Science, and Technology (MEXT, *Monbu-kagaku-shō*) came to supervise the JAEA. It also took over STA's remaining task related to nuclear research. While MEXT was in charge of related research, nuclear policymaking and implementation were firmly in the hands of METI.

The Hashimoto reforms actually presented an opportunity to improve nuclear safety. The nuclear safety bureau inside ANRE was elevated to an agency, the Nuclear and Industrial Safety Agency (NISA, *Genshiryoku-anzen-hoan'in*). However, there was no expansion of competencies vis-à-vis the electric utilities. For example, in-service standards for nuclear power plants—taking into account the wear and tear after what could be decades in operation—were still missing. Once a license was granted, NISA still had no means to force utilities to retrofit and upgrade reactors. Disaster countermeasures, in-use standards, and backfitting remained in the hands of the utilities. NISA's vague legal status left it financially and organizationally dependent on METI. It continued to be staffed by METI with officials dispatched to NISA temporarily, who often lacked expertise in overseeing nuclear facilities. Despite the upgrade on paper, the reform failed to address organizational shortcomings and to give safety regulators more competencies to regulate the industry.

Following the creation of NISA, METI added one more safety organization in 2003. The Japan Nuclear Energy Safety Organization (JNES, *Genshiryoku-anzen-kiban-kikō*) was established to handle operational aspects related to nuclear safety regulation. According to a nuclear safety governance expert (interview, 2015), JNES recruited much of its staff from the industry. As a result, many former industry employees monitored nuclear safety, operated emergency centers at nuclear facilities, supported nuclear emergency exercises at the local level, and managed an emergency response support system. But JNES was not the only administrative agency that took in industry staff.

In the 2000s, staff flow from the industry to the administration grew. Against the background of an increased need for highly specialized staff to oversee nuclear facilities, budgetary concerns in the wake of the Hashimoto reforms, and the depression in the nuclear industry, there was an influx of retired or seconded industry staff. NISA was also partly staffed with technical staff from the industry. The same was the case with the NSC. In 2000, the NSC had seventeen officers on payroll and no technical staff, but in 2007 it had sixty-six officers and forty-one technical staff, most of whom were retired engineers from nuclear manufacturers (Hymans 2011, 182). As the administration took over functions the industry had previously handled in STA, dispatching industry engineers to administrative positions opened the door for continued industry influence over regulation.

By the early 2000s, METI had taken over many competencies in nuclear energy policymaking, implementation, and safety regulation from the Cabinet Office and was *the* central player in the nuclear policy domain. Around the same time, rising oil prices led to an increased interest in nuclear power globally, dubbed the "nuclear renaissance," which succeeded the global nuclear industry downturn of the 1990s. Against this background, nuclear power became ever more central to Japan's energy and economic policy and was adopted as an important instrument for Japan's climate policy, too. As the host of the 1997 climate summit and the name-giver to the 1997 Kyoto Protocol, Japan pledged to reduce its CO_2 emissions by 6 percent by 2012 (compared to 1990 levels). MITI pushed for nuclear power as a means to achieve emissions reductions without burdening the heavy industry too much (Watanabe 2011). At the same time, Japanese manufacturing companies, namely Toshiba, Mitsubishi, and Hitachi, embraced the nuclear renaissance promoted by MITI/METI and moved into the heart of the global nuclear industry. Toshiba increased its stakes in Westinghouse, and Hitachi and Mitsubishi entered tie-ups with GE and Areva, respectively, putting Japan at the nexus of the global nuclear industrial complex.

Hitherto, nuclear power was connected to economic development, both to keep electricity costs down and as an innovative future technology that would bring Japan a "brighter future." In response to the oil shock, expanding the share of nuclear power became part of Japan's energy security strategy of reducing its import dependence. As a low-carbon technology, nuclear power consolidated its stance as the perfect technology to meet the "3Es" of energy policy—energy security, economic efficiency, and environmental friendliness.

In 2006, METI produced a strategy paper to make Japan a "Nuclear State" (*Genshiryoku rikkoku*) (METI 2006a). The idea was embraced by the LDP government under Prime Minister Koizumi in the 2006 "Strategic Energy Plan" (SEP, *Enerugī-kihon-keikaku*).[10] It set a share of nuclear power in the nation's electricity mix of 30 to 40 percent by 2030 and vowed to make "safety issues a priority" (METI 2006b). Interestingly, there was no mention of the electricity market liberalization that had been pursued in the previous decade. As Kikkawa (2012, 115–16) writes, the term disappeared from the 2006 Strategic Energy Plan, because METI followed the nuclear industry's reasoning about more competition suppressing investments into nuclear power. The resulting downturn in nuclear power development would also have an adverse effect on energy security by increasing reliance on imported fuels amid high oil prices. Therefore, the disappearance of retail electricity market liberalization from the plan was a sign of political and administrative support for the nuclear industry.

With plans for a "Nuclear State" and the challenge of local opposition and a more critical public, state support for nuclear power development was ever more crucial. To address the siting issues that emerged in the 1990s, METI intensified the use of the *dengen-sanpō* system even further. It expanded the range of eligible communities to include those planning to host a nuclear power plant, those with one under construction, and those that were simply considering it (Aldrich 2014, 85–86). At the same time, the nuclear village promoted nuclear power as a safe technology that could revive Japan's economy after the prolonged recession of the 1990s, which became known as the "lost decade." As part of this promotion, the nuclear triangle influenced media reporting to portray nuclear power in a favorable light and to omit the risks pointed out by critics. The "nuclear blind spot" in the Japanese press prior to 3.11 (Avenell 2012) was the result of "journalist clubs" (*kisha club*), which contributed to a pro-elite bias in media reporting (Freeman 2000), and large-scale PR campaigns by pronuclear actors (Weiss 2020). Close links between newspapers, governmental administration, and the nuclear industry enabled influence on media reporting about nuclear power via advertising money (Weiß 2019; Honma 2012). Efforts to mold public opinion in favor of nuclear power were eventually successful. Opinion polls from 2011 show that by then a majority of respondents favored nuclear power development once again (Shibata and Tomokiyo 2014, 9). To revive the ailing nuclear power program as part of the nuclear renaissance, the nuclear triangle dropped full market liberalization, expanded the system of subsidies for local host communities, and promulgated the safety myth.

Notwithstanding political and administrative support, there were increasing tensions between the two industry groups. As part of the nuclear renaissance, nuclear power was becoming more attractive for manufacturers as an export technology, while electric utilities worried about the costs of constructing and operating nuclear power plants. In 2006, Chubu Electric Power filed a lawsuit against the manufacturer Hitachi over a faulty reactor turbine that had caused an accident at the Hamaoka nuclear power plant. Dealing with this conflict in open court instead of coming to an agreement between themselves illuminated the emerging conflict between the two sectors. Meanwhile, JAIF—established to facilitate a common position among industry actors—continued to exist, but it was downgraded and no longer acted a consensus-building mechanism (Hymans 2011, 178–79). Instead, each group was represented by their respective federations, electric utilities by Denjiren and plant manufacturers by Keidanren.

METI used its position at the heart of the nuclear triangle to promote and support nuclear power development on many fronts. Beneath the surface of overt administrative and political support, however, policy implementation was encountering challenges. As figure 1.1 shows, the expansion of nuclear power generation capacity had come to a halt and even showed decline in the 2000s. Nuclear industry groups, plant manufacturers, and electric utilities were not acting in concert anymore as rising costs made nuclear power a less profitable business for

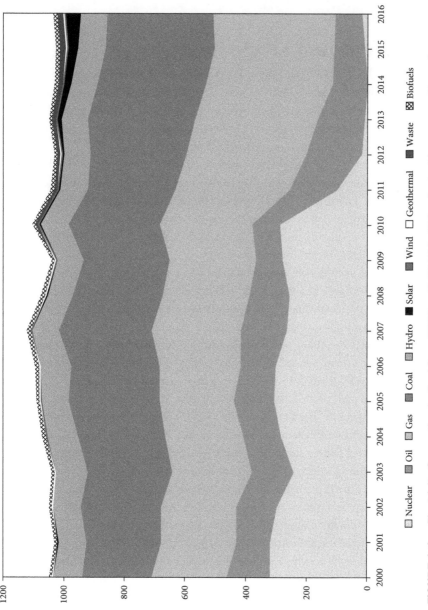

FIGURE 1.1. Electricity Generation by Fuel from 2000 until 2016 in TWh, stacked by source with nuclear power at the bottom, created by the author based on data from the International Energy Agency

utilities. Against this background, what seemed like a very strong METI-centered nuclear program was in fact an attempt to revive Japan's struggling nuclear power program.

Attempts to Improve Nuclear Safety

In line with the 2006 Strategic Energy Plan's promise to improve safety, the safety standards issued by the AEC in 1978 were revised in 2006. Changes included an expanded definition of active fault lines to include seismic activity going back 120,000–130,000 years as opposed to only 50,000 years previously. Mentioned for the first time were tsunamis as events accompanying earthquakes. But there were no concrete tsunami countermeasures required. The 2006 safety standards omitted the scientific progress made in the study of tsunamis, which Shiroyama described as a "failure of interdisciplinary communication" (2015, 289–90). New safety standards applied to the construction of new nuclear reactors and not to refurbishment of existing reactors. Therefore, the 2006 safety standards applied to only a portion of Japan's nuclear reactors.

When NISA and the NSC attempted to implement stronger safety measures at existing nuclear power plants, their efforts were met with resistance. In 2009, the NSC and NISA calculated new basic ground movement standards for nuclear reactors to withstand in case of an earthquake. The backfit investigations they conducted concluded many plants to already be equipped to withstand stronger seismic tremors. However, all plants were restarted before investigations were even concluded (Tateishi 2015, 97), showing the industry's unwillingness to idle nuclear power plants for safety improvements. In 2008, a tsunami simulation showed the possibility that a tsunami caused by an earthquake could breach the Fukushima Daiichi NPP's tsunami protection wall. But TEPCO ignored the recommendations of NISA and the NSC that it put in place a higher protection wall at the Fukushima Daiichi NPP, as a trial against three TEPCO executives revealed in 2018 (The Diplomat 2018). Had a higher seawall been installed, the Fukushima nuclear accident most likely would have been on a much smaller scale. These episodes illustrate the relative powerlessness of NISA and the NSC in the face of an undiscerning industry that opposed costly safety measures.

After the string of accidents in the 1990s and an international recommendation to improve safety governance, METI looked into ways to make improvements. A review of Japan's nuclear safety administration, carried out by the International Atomic Energy Agency (IAEA), concluded in 2007 that the role played by different actors within nuclear governance, NISA and the NSC in particular, should be clarified. It further recommended that NISA's supposed independence from ANRE should be more clearly reflected in the legislation and pointed out the need for a strategy to secure skilled staff for the nuclear safety administration (Caruso 2012). METI set up the Nuclear Legislation Study Group, an advisory group to make reform recommendations.

Running from 2007 until 2010, the study group discussed ways to improve technical safety regulations and the legal framework of the safety regulator. The interim report, published in July 2009, recommended improving technical regulation and changing the regulators' institutional design to make them more independent. The report by the study group was handed over to NISA and the NSC and presented before the AEC. According to a study group member, some in the AEC were interested in changing the agency setup, but there was not enough momentum to push for it (interview with Nuclear Safety Governance Expert, 2015). A former AEC member recalled the presentation in which the Nuclear Legislation Study Group presented a comprehensive change in the legal framework for nuclear safety regulation. It would have eliminated the duplication of electricity industry regulation (NISA) and nuclear safety regulation (NSC). He recalled that the five AEC members liked the recommendations but were aware of the opposition they would encounter trying to implement such a reform (interview with Former AEC Official, 2016). Ultimately, the expert group's recommendations for administrative and technical changes ended up in the drawer instead of leading to changes in the system. Not long before 3.11, efforts to address the shortcomings in Japan's safety regulation system were stymied by a lack of momentum to undertake reforms unpopular with the nuclear industry.

The LDP losing power in 2009 could have been a game changer for nuclear policy, because political support by successive LDP governments was an important part of the nuclear triangle. In the 2009 elections, the Democratic Party of Japan (DPJ, *minshutō*) won with campaign slogans such as "political leadership" (*seiji shudō*) and abolishing the LDP's "ancient regime." The DPJ, however, left the LDP's pronuclear policy untouched. One reason was the party's support base, which consisted of labor unions related to the power companies and electronic product manufacturers (Hasegawa 2012, 88). Another reason was a lack of alternative policy ideas. As Zakowski put it: "The DPJ was not much more than a patchwork of different groups, the refinement of the grand design for reforms quickly became a matter of power struggles rather than an object of intraparty consensus" (2015, 12). With a support base close to the nuclear industry and no alternative energy policy vision, the DPJ government adopted the 2010 "New Energy Strategy" with the intention to build nine new or additional nuclear reactors by 2020, and more than fourteen by 2030. This was supposed to increase the ratio of nuclear power in the energy mix to 53 percent by 2030 (METI 2010, 4). Rather than unravel the nuclear energy system, the DPJ pragmatically embraced nuclear power—at least until 3.11.

The Nuclear Village and Regulatory Capture

This chapter set out to trace the institutional arrangement that buttressed half a century of nuclear power expansion. Basically, Japan's nuclear energy policy domain rested on two prototypical pillars. The nuclear iron triangle consisted of the LDP, METI, and the nuclear industry, that is, big manufacturing companies and electric utilities. Its institutions structured the political conflict in a way that limited decision-making power to pronuclear actors and largely kept the public out and critics silent. Furthermore, they heavily promoted supporting ideas, such as the importance of nuclear power for economic growth and energy security as well as the nuclear safety myth, to depoliticize nuclear power and to sustain public support for its development.

The institutional arrangement of the nuclear iron triangle created a vested interest structure in favor of expanding nuclear power further. For political decision makers and METI, expanding nuclear power was a central means to keep electricity prices low in support of the economy, to increase Japan's energy security by lowering import dependency, and to reduce greenhouse gases and other emissions as a way to protect environment and climate. The more that the LDP and METI invested into nuclear power to achieve these goals, the more that successful implementation mattered. METI officials also profited from retirement positions in the industry they used to oversee, while the LDP received considerable financial campaign support from nuclear industry actors. Furthermore, the more tax money was used for R&D, the promotion of civilian nuclear power, and the necessary technical infrastructure, the more expensive a radical change in energy policy path became. If Japan had given up nuclear power, all those investments into a "bright future" would have been lost. With time, increasing returns and vested interests of the nuclear village increased the incentive to defend the institutional arrangement from outside challengers.

Nuclear policy implementation increasingly depended on state support for nuclear power projects. Plant manufacturers bet on nuclear power as a future technology after World War II and, again, amid the nuclear renaissance of the 2000s. They needed a market for their nuclear reactors, for example, by expanding nuclear power within Japan and by exporting reactor technology. To enable this, Japanese state support included the financial support scheme *dengen-sanpō*, a system of regional energy monopolies, and strategic information management. The importance of state support became apparent after the 1990s witnessed nuclear power expansion almost coming to a halt due to local opposition. To address policy implementation challenges, political support became even more pronounced as part of the nuclear renaissance. State support was paramount to a conducive business environment for nuclear power.

Within the nuclear triangle, industry actors enjoyed a powerful position that allowed them to influence policy decision-making and safety regulation. Early attempts at state control were met with fierce opposition by the nuclear industry. As the winners of this initial struggle, the nuclear industry, and electric utilities in particular, henceforth defeated all attempts to subject them to binding state regulation, such as accident countermeasures. Safety governance reforms in the 1970s and early 2000s made it appear as if the nuclear iron triangle reacted to safety concerns. However, the reforms mainly transferred power from the Cabinet Office to METI but left industry competencies untouched. With the Hashimoto reforms, Japan's nuclear program became fully centered on METI. Out of four administrative bodies responsible for nuclear policy and regulation, the more powerful ones were METI's ANRE (energy policy) and NISA (nuclear safety). Their weaker counterparts at the Cabinet Office were the AEC (nuclear policy framework) and NSC (safety double-checking). At the center of nuclear policy and safety, METI continued the system of integrated promotion and safety regulation initially established under what was then the Prime Minister's Office. Pre-3.11 reforms were mere window dressing because they failed to address the institutional features that enabled industry influence over safety regulation.

The Japanese reality of a sense of shared interest and intertwined careers contributed to safety agencies too weak to impose safety measures. The safety agency NISA (and its predecessor nuclear safety division within ANRE) depended on METI for its finances and human resources. NISA was mainly staffed with METI officials, who transferred in and out of the safety regulator every few years with few being experts in nuclear safety regulation. Ubiquitous staff rotation within ministries and with agencies at the Cabinet Office allowed METI to exercise influence over the NSC, too. The practice of many METI officials later retiring to the nuclear industry created no incentive for them to strictly regulate the industry they sought to retire to. Furthermore, both NISA and the NSC relied on industry expertise. In the 2000s, it became a widespread practice that industry experts took up temporary positions as safety regulators to fill a gap in state expertise. This intense revolving-door phenomenon clearly furthered the industry's ability to influence safety regulation and limited incentives for officials to push for better safety governance. The organizational features of integrated nuclear safety and promotion left regulators weak and prone to capture, most notably due to a lack of in-house expertise and regular staff rotation between nuclear promoters, regulators, and the industry.

These two features, weak regulating agencies and an industry bias, resembled a classic example of regulatory capture where "a regulated industry is able to control *decisions* made about that industry by regulators and/or *performances* by regulators related to the industry." (Mitnick 2011, 35). Some criticize the idea of capture in the cooperative Japanese context due to a lack of evidence of the industry taking over the regulator (Scalise 2015). In the case of Japan, capture was not the result of a takeover but of a long-running intense exchange between the state and industry. Early on, the industry became a partner in Japan's nuclear power program, including policymaking and setting safety standards. As accidents revealed the risks of nuclear power, this setup became flawed, that is, biased in the industry interest. This was clearly visible from the window-dressing reforms that took place after nuclear accidents. Even though the government reformed the nuclear safety administration in the 1970s and 2000s in the name of public safety, a clear public interest, it never granted safety regulators the legal framework and organizational autonomy to impose necessary safety standards on the nuclear industry due to industry opposition. Mere years before 3.11, calls by the IAEA and the Nuclear Legislation Study Group to increase regulatory independence and to improve accident countermeasures fizzled out due to anticipated industry opposition. By that time, Japan's nuclear safety governance had clearly been captured.

Nuclear accidents inside and outside of Japan challenged the positive narrative created by agencies promoting nuclear power and the safety myth promulgated by the industry. Initially intended to garner public support for civilian nuclear power development after the traumatic experience of two nuclear bombings, strategic information management eventually turned into a system to conceal the risks of civilian nuclear power. A bias in press coverage, education on the benefits of nuclear power, decision-making behind closed doors, and a lack of transparency kept the public in the dark even after accidents around the world and in Japan showed the risks involved. Hence, the nuclear iron triangle used its massive resources and power vis-à-vis outsiders to defend their position by way of strategically managing information to foster a safety myth and to shape public opinion in favor of nuclear power while simultaneously suppressing critical voices in safety regulation and policymaking.

Those unsatisfied with nuclear power policy and regulation lacked access to decision-making. Apart from elections, the political system mainly offered entry points at the local level. Antinuclear protests grew over time, centered on NIMBY protests against individual plants. If they tried to use courts to make their voices heard in the political conflict over nuclear power, Japanese courts ruled in favor of the state. This was in line with Ramseyer's (2019) argument that courts in Japan were under the influence of the LDP, leading them to rule in favor of the government on contentious issues. The Japanese judiciary dismissing cases against nuclear power plants or ruling in favor of such projects buttressed the power of the nuclear iron triangle.

Over time, the costs of nuclear power development emerged as a possible Achilles' heel of the nuclear power system. The nuclear iron triangle was crucially based on nuclear power as a profitable business choice. This was exemplified by the nuclear industry pulling out of the fast breeder reactor Monju project after costs skyrocketed. Similarly, the biggest electric utility, TEPCO, postponed expensive safety measures—with the tragic consequence of a large nuclear accident—in order to keep costs down. More generally, the nuclear industry successfully lobbied to keep technical costs down by preventing a regulatory governance mode with stringent safety standards at higher technical costs. Another kind of costs stemmed from local protests. Social acceptance costs refer to the additional costs caused by citizens turning against a nuclear power project. These include cost overruns for electric utilities, if a project is stalled, but also legal costs of fighting off lawsuits. To avoid acceptance costs, utilities shifted from siting nuclear power plants to expanding existing ones in the 1990s. Court rulings in favor of the state and utilities prevented the total shutdown of nuclear power projects, thus helping to limit social acceptance costs. The success of Japan's nuclear power program crucially depended on keeping technical and social acceptance costs low.

In sum, this chapter has traced the pre-3.11 system of regulatory capture, strategic information management, and overt state support fostered by the nuclear iron triangle. A lot of this goes back to decisions made when Japan's nuclear program was put in place. The stability of Japan's nuclear power program rested on institutions within the nuclear power domain that empowered the iron triangle. Path dependency was further reinforced by institutional arrangements in the energy domain (energy monopolies), bureaucracy (*amakudari*), academia (close industry ties through funding), Japanese media (journalist clubs and advertising money), and judiciary (political influence via judiciary appointments). Meanwhile the costs of building and operating nuclear power plants emerged as a possible Achilles' heel of the system. In the language of costs often used by historical institutionalist scholars, the nuclear village kept technical costs down at the expense of public safety and limited social acceptance costs by largely keeping the public and critics out while simultaneously becoming more and more invested in their own arrangement over time. As long as those costs remained low, the increasing returns of nuclear power development outweighed them and created an incentive to stay the course.

3.11 AS AN OPPORTUNITY FOR CHANGE

The earthquake, tsunami, and nuclear accident that assailed Fukushima, Japan, on March 11, 2011, cast a harsh spotlight on nuclear safety. Almost immediately, there were initiatives to strengthen safety measures at nuclear power plants abroad. In late March, the EU began its stress test, a review of nuclear power plant safety in EU member states. In April, representatives from different countries gathered at the International Atomic Energy Agency (IAEA) to discuss possible nuclear safety improvements. The consultation process resulted in the IAEA adopting the "Action Plan on Nuclear Safety" in September 2011 in order to improve nuclear safety and emergency preparedness worldwide (IAEA 2012). Such international actions put pressure on the Japanese government to act swiftly and decisively. Furthermore, many countries distrusted the Japanese government's reassurances that Tokyo was safe and that the accident was under control. The United Kingdom, France, Italy, and Australia relocated their embassies from Tokyo to cities further south and thus further away from the site of the accident. Other countries advised their citizens to leave Japan altogether (Guardian 2011b). The international community responded swiftly to improve nuclear safety and showed signs of distrust in Japanese safety governance and handling of the accident.

The nuclear accident at the Fukushima Daiichi nuclear power plant shook Japan to its core. Shock events, such as accidents, disasters, and economic crises, throw a spotlight on deficiencies of existing policies and institutions and induce calls for improvements (Kingdon 1984; Baumgartner and Jones 1993; Blyth 2002; Hogan and Doyle 2007; Birkland 2007; Capoccia and Kelemen 2007; Balleisen et al. 2017). For decades, it was said that an accident such as those at Three Mile

Island in the United States or at Chernobyl in what was then the Soviet Union was impossible in Japan due to the superiority of Japanese nuclear power technology, Japan's so-called safety myth (discussed in chapter 1). Unsurprisingly, 3.11 stimulated calls for changes in nuclear safety governance and for a nuclear phaseout.

Reform outcomes after 3.11 are emblematic of how shock events translate into substantial changes in some instances but not in others. Substantial safety governance reforms resulted in the establishment of an independent safety regulator, but a phaseout was ultimately rejected. Hence, there was continuation in nuclear power policy, albeit based on revamped nuclear safety institutions. This chapter compares how the two reform initiatives unfolded in order to understand how agents of change pushing for safety governance changes overcame vested interests.

There are three types of agents of change. First, policy entrepreneurs are those who produce policy-related ideas and have access to the political process, for example bureaucrats, some academics, and interest groups. Second, political entrepreneurs denote figureheads who introduce a new policy idea, mainly politicians, and often opposition leaders. Then, there are outside influences, which can be domestic actors—the media, the public, and social movements— or international actors, including international organizations or foreign models. While the third type of agents of change is located outside of the political decision-making process, they can become influential (Hogan and Doyle 2007).

There are two approaches to explain whether or not agents of change are successful in pushing for reforms. One explanation focuses on power relations between agents of change and vested interests seeking to prevent changes to institutions they profit from. If agents of change are able to overcome the resistance of vested interests and veto players, they are able to steer the reform process toward substantial change (Capoccia and Kelemen 2007; Capoccia 2016b; Balleisen et al. 2017). Another explanation draws on ideas as a source of change. To begin with, the existence of alternative ideas matters (Blyth 2002). If different agents of change consolidate around such an idea for an alternative institutional design, there will most likely be a change in that direction (Hogan and Doyle 2007; Birkland 2007; Baumgartner and Jones 1993). Two possible explanations emerge from this: 1) agents of change are successful if they have the power to push through their preferred alternative; 2) the more that different agents of change consolidate around one alternative idea and collectively push for it, the more likely change in that direction becomes.

To understand the different outcomes produced by post-3.11 reform initiatives, it is necessary to reconstruct the decision-making process that led to safety governance changes and the resolution to retain nuclear power. Using process tracing to reconstruct each step of the decision-making process, it is possible to identify influential decisions and available options at the time (Capoccia and Kelemen 2007). How influential a decision was depends on how different the reform outcome would have been, had a discarded alternative been chosen (Capoccia 2016a). The difference between a chosen and a discarded option is revealed by a counterfactual analysis based on established regularities, theories, or principles in existing literature (Fearon 1991) and a researcher's background knowledge (Collier 2011).

Once it is apparent *which* decisions were important, the question is *who* was behind it, and *why* they advocated this one rather than an alternative. Actors' motivations can be inferred through framing analysis. Framing analysis looks at how actors attribute meaning to events and, thus, arrange experiences in a meaningful manner to use it as a guide for action. First, diagnostic framing identifies the source of the problem at hand. It seeks out the culprit who is to blame for the problem (Stone 1989), which is a crucial element of the narratives different actors seek to construct in the aftermath of a shock event (Birkland 2007; Blyth 2002). Second, prognostic framing articulates a solution to rectify the problem. Diagnostic and prognostic framing are closely connected, because what one understands to be the problem affects what constitutes a suitable solution. Lastly, motivational framing serves to mobilize others by providing a rationale for engaging in collective action (Benford and Snow 2000). Framing analysis is useful here to distinguish the positions that different actors took in the post-3.11 reform struggle. The main actors in political processes are not individual but rather collective actors, such as a political party or a government, which take a position as a result of internal negotiations. To identify their motivations, the approach here is to look at official statements by the collective actor as well as individual statements by the persons handling reform matters on behalf of the collective actor.

This chapter seeks to understand why 3.11 led to fundamental change in Japan's nuclear safety governance institutions but no nuclear phaseout policy, and what this reveals about mechanisms for change in the aftermath of a shock event. To do so, it traces which decisions were influential, who was behind them, and why it was adopted rather than an alternative one.

Public Outcry Following 3.11

The shock about 3.11, and the nuclear accident in particular, shifted public opinion on nuclear power towards a more critical stance. Public opinion polls by both the center-left *Asahi Shimbun* and the conservative pronuclear *Yomiuri Shimbun* asked participants about their preferred future nuclear power policy. As shown in figure 2.1, the Yomiuri poll started out with higher support numbers, but both show a clear increase in the percentage of people favoring a decrease or even a

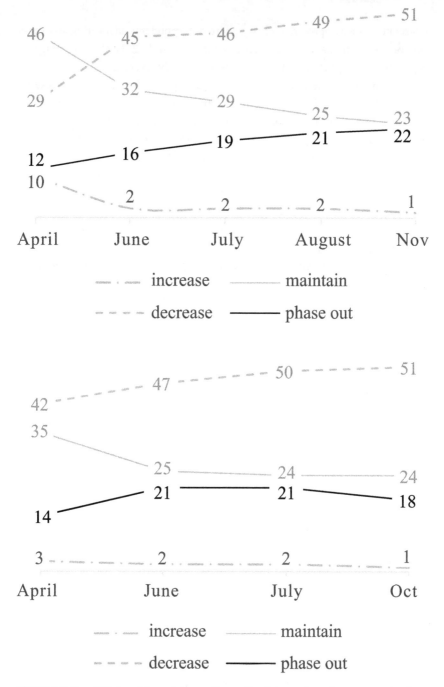

FIGURE 2.1. 2011 public opinion polls on the future of nuclear energy by the *Yomiuri Shimbun* newspaper (top) and the *Asahi Shimbun* newspaper (bottom), shown in percent (%), created by the author based on data from Shibata and Tomokiyo (2014)

phaseout of nuclear power, while the share of respondents advocating an expansion or the maintenance of the status quo decreased. In 2011, the percentages of supporters and opponents of nuclear power basically reversed.

Concomitant to the shift in public opinion, large public demonstrations swept through the country. Demonstrations began when fifteen thousand people demonstrated in Koenji, Tokyo, on April 10, 2011, and again in Shibuya, Tokyo, on May 7, according to organizers. Protests began to sweep the country with approximately 140 demonstrations held nationwide on June 11. The movement reached its 2011 peak with sixty-five thousand people gathering in Meiji Park, Tokyo, on September 19, according to organizers. Participants included famous author and Nobel laureate Ōe Kenzaburo. Usually subsumed under the term "antinuclear demonstrations," participants focused on different issues connected to nuclear power. One major group were women, particularly from Fukushima Prefecture, who came with their families to speak about their ordeal in the weeks following the accident and concerns about their children's health after exposure to radiation. Another group, mostly elderly people, protested against nuclear weapons and in favor of keeping Japan's pacifist constitution. A third group were mainly young people protesting against the use of nuclear power as an energy source (Hasegawa 2014). In reaction to 3.11, the largest public demonstrations since the 1960s (Slater 2011) targeted the government's crisis management as well as use of nuclear power technologies.

The Kan Cabinet's Reform Initiatives

The Kan government had to deal with three disasters that occurred on March 11: a magnitude 9 earthquake, a fifteen-meter-high tsunami devastating large stretches of Japan's northeastern coast, and the unfolding nuclear accident at the Fukushima Daiichi nuclear power plant. Such an unprecedented event can best be described as confusingly complex. In such a complex situation, the government depends on experts in respective agencies to provide advice and to help manage the situation. The Nuclear and Industrial Safety Agency (NISA), however, initially failed to gauge the severity of the accident. It had to revise its March 12 assessment as a level 4 "accident with local consequences" (IAEA 2011b) to a level 7 "major accident" one month later (IAEA 2011a). Furthermore, a delay in reporting radiation monitoring data back to the government from the Ministry of Education, Science, and Technology (MEXT), which was in charge of the radiation monitoring system SPEEDI, resulted in late evacuations and in a step-by-step process that exposed evacuees to higher radiation levels. The accident response garnered Kan much criticism, with some faulting his response as too little too late while others lamented too much micromanagement (Samuels 2013; Lukner and Sakaki 2011).

The flawed accident response undermined the legitimacy of the nuclear safety administration in the eyes of the Kan government. After the accident, Prime Minister Kan requested NISA, the Nuclear Safety Commission (NSC), and TEPCO to send people to the Prime Minister's Office to provide expertise and counsel (Kan 2012, 60) in line with provisions in the Nuclear Disaster Countermeasures Special Measures Law from 1999. While NSC head Madarame Haruki immediately began providing advice, Kan perceived TEPCO and NISA to struggle with their assigned role. The exchange of crucial information with TEPCO proved arduous, prompting Kan to set up an emergency response headquarters at TEPCO and to place his adviser Hosono Gōshi there to ensure better communication (Kan 2012, 33). NISA was supposed to staff and run the Nuclear Disaster Countermeasures Headquarters, but the NISA head Terasaka Nobuaki was an economist from METI who knew no more about nuclear disaster management than the government asking for advice (Kan 2012, 17). Not that there were no experts within NISA, but it took a few days until Yasui Masaya, head of the department for energy saving and new energy sources,[1] assumed the position intended for the head of NISA. This crisis management experience left the DPJ cabinet collectively skeptical of METI and NISA and shaped their idea of what nuclear safety governance reforms should take place.

According to Kan, the source of the problem—the apparent lack of expertise within NISA—stemmed from METI's influence over NISA via temporarily placing officials there as part of staff rotation within Japanese ministries (Kan 2012, 63–64). The distrust harbored by Kan and his cabinet towards METI and NISA manifested in the decision to put the Ministry of Environment (MOE) in charge of decontamination efforts, even though the MOE at the time lacked radiation experts (interview with Former AEC Official, 2015). Setting the basic direction of nuclear safety reforms, Kan gained METI Minister Kaieda Banri's consent to separating NISA from METI. He also appointed his adviser Hosono as State Minister for the Restoration from and Prevention of Nuclear Accidents (henceforth Minister for Nuclear Affairs) and put him in charge of making reform arrangements (Kan 2012, 169–70). By preempting a METI veto to his reform idea and by putting his trusted adviser Hosono in charge of reform preparations, Kan led the effort to reform nuclear safety governance. Since previous nuclear safety governance reforms under LDP rule had integrated nuclear safety regulation more and more within METI, the decision to separate NISA from METI was a far-reaching first decision.

Hosono and his team of about ten people were in charge of drafting a decision for a new safety regulator. Members included officials from the MOE, METI, and the police agency, with the most important person being Morimoto Hideka from the MOE (interview with Former Administrative Vice Minister, 2015). Morimoto spent most of 2011 and 2012 working on the reforms at the Cabinet Secretariat and later assumed an executive position in the new agency. As part of the preparations, the team contacted Shiroyama Hideaki, a member of METI's Nuclear Legislation study group from 2007 to 2010, whose recommendations had ended in the drawer (interview with Nuclear Safety Governance Expert, 2015). Shiroyama's writings on the need for an independent unified safety governance institution with greater in-house expertise (Shiroyama 2015, 2012, 2011, 2013) make it likely that he stressed these points.

Early on, the DPJ reform proposal stressed greater independence, to be achieved by separating nuclear power regulation and promotion, and stronger expertise within the new agency. On August 5, Hosono presented five principles: 1) separation of regulation and promotion functions; 2) unification of relevant functions regarding nuclear safety regulations; 3) strengthening of the crisis management functions; 4) reform of organizational culture, and securing and training talented experts; and 5) strengthening of nuclear safety regulations (Cabinet Secretariat 2011c). Cabinet members agreed on replacing NISA with a new agency staffed with between five hundred and six hundred people (Nihon Keizai Shimbun 2011, 3). The location of the new regulatory agency, however, was a point of contestation at first. A group around Chief Cabinet Secretary Edano Yukio and Vice Cabinet Secretary Sengoku Yoshito advocated for the Cabinet Office, as they harbored doubts about the MOE's ability to handle crisis response management, particularly to mediate between different ministries and agencies. In contrast, Kan and Hosono presented the MOE as a suitable option, due to its experience in handling environmental pollution issues such as air pollution and because of a lack of ties with the electricity industry (Nihon Keizai Shimbun 2011, 3; Asahi Shimbun 2011a, 3). When the cabinet adopted the five reform principles on August 12, it agreed on affiliating the new organization with the MOE (Cabinet Secretariat 2011d).

During reform preparations at the cabinet, others weighed in on the reform proposal. It was under discussion to move the radiation monitoring department out of MEXT and into the new regulating body. This prompted MEXT to protest through having its executives visit several Diet members to dissuade them from the idea (Asahi Shimbun 2011a). In the debate about where the new regulator should be placed, METI opposed the idea of placing it at the Cabinet Office by referring to the fact that a collection of staff from different ministries work at the Cabinet Office, which makes it a place for interministerial power struggles. A METI official likened an affiliation of the nuclear safety agency with the cabinet with putting "a koi on a chopping board." Rather, high-ranking METI officials supported the MOE option (Asahi Shimbun 2011b, 4). It may seem surprising that there were no signs of METI opposing the reform efforts the same way as MEXT did, but METI Minister Banri agreeing to remove nuclear safety regulation from METI had largely eliminated its power to prevent such a move.

In the meantime, the question arose how to ensure the immediate safety of nuclear power plants. NISA and the NSC worked out emergency safety measures for nuclear plants after the accident. In April, the electric utilities association Denjiren assured the public that respective emergency safety measures were being taken at nuclear power plants (Denjiren 2011b). Similarly, METI issued reassurances to the public that operational safety had been confirmed (Kingston 2014, 115). In contrast, the AEC considered some of the emergency planning provisions to be insufficient. To address these concerns, the AEC called for a stress test at nuclear power plants. METI, however, opposed the idea, because it was wary of shutting down nuclear power plants without clearly defined conditions for restarts (interview with Former AEC Official, 2015), but the stress test was later ordered by the Kan government anyway. Once ordered, the Denjiren chairman pledged to conduct the stress test with the utmost care in order to regain the trust of the people (Denjiren 2011a) but refrained from commenting on the reform proposal per se in any press conferences in 2011. While neither METI nor Denjiren openly opposed reform plans, they rushed to assure the public—and the government— that nuclear power plants were safe in their hands, implying no need for a stress test—or a reform of the nuclear safety administration.

The politician Shiozaki Yasuhisa, the LDP's leader on nuclear safety reforms, early on criticized the government's lack of attention to making the new regulating agency as legally independent as possible by establishing it as an independent regulatory commission based on Article 3 of the National Government Organization Act. Shiozaki regarded this as the most appropriate legal status because it was in line with the independence criteria spelled out by the IAEA and the example set by nuclear regulation agencies in other countries, such as the US Nuclear Regulatory Commission (NRC) (Shiozaki 2011a). In his later book Shiozaki repeated his position that Japan should follow the "IAEA Safety Standards" as "an international charter of nuclear regulation" (Shiozaki 2012, 27). Shiozaki also stressed agency independence in a speech at the Stimson Center in Washington on May 6, 2011, where he declared:

> Japan needs to create a totally independent agency staffed with professional who can monitor and regulate the Japanese nuclear industry in a transparent manner. The so-called "iron triangle" of politics, bureaucracy, and business must be dismantled to restore confidence both domestically and internationally. (Shiozaki 2011c)

This statement clearly indicates that Japan's nuclear safety reforms had in fact two key audiences: a domestic and an international one. Overall, Shiozaki's early reform statements stressed adherence to international norms in the form of an independent nuclear regulator and foreshadowed the LDP's 2012 election strategy of attacking the DPJ and its reform proposal as lacking in independence provisions.

On August 15, 2011, mere days before Kan's resignation, the Kan Cabinet decided on the "Basic Policy on the Reform of an Organization in Charge of Nuclear Safety Regulation," which resolved to separate nuclear regulation and promotion and to integrate the different nuclear safety functions in an agency affiliated with the MOE. Along with fusing NISA's nuclear safety functions and the NSC, the cabinet decision determined to integrate MEXT's radiation monitoring system SPEEDI into the new agency (Cabinet Secretariat 2011c). A few days prior, Kan had announced the measure of supplementing the proposed unified regulator with a nuclear safety policy council staffed with nuclear safety specialists appointed by politicians (Yomiuri Shimbun 2011, 2). The MOE was gearing up to take over preparations for a new regulatory agency and was looking to establish a team in charge of setting up the new agency (Asahi Shimbun 2011a). Keeping the bill-drafting process close, however, the Kan Cabinet established the Task Force for the Reform of Nuclear Safety Regulations and Organizations (Task Force, *Genshiryoku-anzen-kisei-soshiki-nado-kaikaku-junbishitsu*) at the Cabinet Secretariat and, again, put Hosono in charge (Cabinet Secretariat 2011a).

Concomitant with safety governance reforms, Kan also pushed for a nuclear phaseout. A nuclear phaseout would have overturned half a century of nuclear power promotion in Japan. It also stood in stark contrast to the nuclear policy goals adopted by METI and the DPJ government under Prime Minister Hatoyama in 2010, which called for an increase in the share of nuclear power to at least 50 percent by 2030 (METI 2010). This development revealed how deeply divided the DPJ was over nuclear power (Watanabe 2013). Within the cabinet, METI Minister Banri and Chief Cabinet Secretary Edano favored a phaseout. Hosono was not opposed, but he did not think it could be done at once. Rather, it required a long-term roadmap, also to ensure that Japan retained skilled labor and necessary technological expertise (Hosono 2013, 201). DPJ government members regarded safety governance reforms and nuclear energy policy reforms as two completely different issues. The former represented a short-term administrative restructuring while the latter was a long-term strategic decision about Japan's energy policy. While everybody supported restructuring the nuclear safety administration due to the crisis management experience (interview with DPJ Government Adviser, 2015), opinions on Japan's future energy policy differed widely.

Without a consensus in the cabinet, and with little government consultation as METI Minister Banri later lamented (Zakowski 2015, 151), Kan announced his nuclear phaseout intentions:

> With regard to Japan's future nuclear power policy, we should aim to achieve a society that is not dependent on nuclear power. In other words, we should reduce our dependence on nuclear power in a planned and gradual manner. (Prime Minister's Office 2011)

His announcement was met with loud opposition by business interests (Vivoda and Graetz 2014, 12). Despite the criticism, Kan's successor Noda Yoshihiko committed to a gradual long-term nuclear phaseout in his inaugural address (Tabuchi 2011). Hence, Kan succeeded in putting the idea on the political agenda without a Cabinet Decision.

The change of heart Kan experienced following 3.11 made him an advocate of energy policy and regulation changes. During the months until his resignation in August, he put his political weight behind a nuclear safety reform, a nuclear phaseout, and the introduction of a feed-in tariff (FIT) to support renewable energy expansion. He conditioned his resignation, demanded loudly by the LDP opposition, on support for passing a FIT bill. Kan's resignation marked the end of the first phase of the reform process with two important decisions made: reforming the nuclear safety administration by moving it out of METI's sphere of influence and considering options to reduce Japan's dependence on nuclear power.

The Noda Cabinet's Policy Drafts

Noda continued the energy policy and regulation changes initiated by the Kan government. He appointed Hosono as Minister of the Environment in addition to his position as State Minister for Nuclear Affairs and appointed the former Chief Cabinet Secretary Edano as Economy Minister. Edano was in charge of the Energy and Environment Council, established in the cabinet to deliberate Japan's future energy policy. Even after Noda took over from Kan, both reform initiatives were pursued under the auspices of the cabinet.

Beginning with Hosono's task force, it sought input from the Advisory Committee for Prevention of a Nuclear Accident (Advisory Committee, *Genshiryoku-jiko-saihatsubōshi-komonkaigi*), which held multiple meetings between October and December 2011 in order to formulate recommendations for Hosono's task force. Via the advisory committee, the DPJ government invited a group of mostly scholars and academics,[2] including critics of nuclear power such as Iida Tetsunari,[3] to be policy entrepreneurs in the legislative process.

In its final report (Cabinet Secretariat 2011b), the advisory committee stated that the problem at hand—overconfidence by the government in extant licensing and safety measures leading to an accident and a loss of public trust in safety measures—required a swift revision of nuclear safety regulation. The report regarded the separation of the promotion and regulation of nuclear energy to be appropriate to achieve independence—the main aim of reforming nuclear governance—and to be in line with previous IAEA recommendations to Japan. To achieve institutional independence, the report recommended the establishment of an independent administrative committee based on Article 3 of the National Government Organization Act. Furthermore, some members maintained that a "no-return rule" should be put in place for staff from related ministries and private utilities. Adding to the five principles adopted by the government, the advisory committee stressed that transparency and internationality were imperative in order to ensure both internal and external accountability and to win back citizens' trust. The advisory committee backed the government's reform direction and recommended even more independence and accountability.

After consulting Japanese experts, the DPJ government sought input and approval from international experts. The Cabinet Secretariat hosted an International Workshop on Nuclear Safety Regulation on January 18, 2012, with the aim of receiving input on the draft bill from representatives from the International Atomic Energy Agency (IAEA), the Nuclear Energy Agency in the Organization for Economic Co-operation and Development (OECD/NEA) and nuclear safety regulation organizations in France, the Republic of Korea, the United Kingdom and the United States (Cabinet Secretariat 2012d, 1). Hosono's task force presented plans for the creation of an external agency of the MOE, staffed with about five hundred people and a budget of ¥50 million. Despite being part of the MOE as an Article 8 agency, it was conceptualized to control its own budget and human resources in order to ensure its independence. The agency was to be headed by a nuclear safety investigation committee subject to parliamentary approval. Regulatory decision-making was supposed to be in the hands of the heads of the safety agency as opposed to the Minister of the Environment. (Cabinet Secretariat 2012c, 3). This workshop essentially invited international actors to be policy entrepreneurs in the legislative process as well.

On behalf of the IAEA, the special coordinator of the IAEA Nuclear Safety Action Plan initiated in the wake of the Fukushima accident, Caruso Gustavo, presented lessons learned and related recommendations. The main lesson learned from Integrated Regulatory Review Service (IRRS) missions in different countries since 2006 was the importance of regulatory independence (Cabinet Secretariat 2012d, 3). For his recommendations, he drew on three findings by the IRRS mission to Japan in 2007. First, the role played by different actors within nuclear governance should be clarified. Secondly, the safety agency's independence from ANRE should be more clearly reflected in the legislation. Thirdly, it pointed out that a strategy to secure skilled staff for the future was needed (Caruso 2012). Jointly, workshop participants came to the conclusion that

> Japan's reform of nuclear safety regulatory organization can be recognized as a great progress in terms of ensuring de jure independence due to clear separation from any undue pressure from interested parties including nuclear promotion bodies. In creating the new organization, it is important that the culture of the new regulatory body also reflects de facto independence by the quality of its actions and regulatory decisions. (Cabinet Secretariat 2012d, 5)

While international workshop participants welcomed the government's reform proposal, they called on the task force to ensure de jure and de facto independence of the new agency. This amounted to a call for an independent Article 3 commission organization and attention to ensuring independence in the regulatory agency's daily business.

Not long after the workshop, the Noda Cabinet adopted the draft bill formulated by the task force, which stipulated integrating functions related to nuclear safety, security, and safeguarding as well as disaster countermeasures into one regulatory body. Concretely, it proposed to fuse NISA's nuclear safety functions, the NSC, MEXT's nuclear research functions, and the AEC's nuclear security functions into the new agency. It did not provide for an incorporation of JNES into the new agency. Rather, JNES was to continue handling the operational aspects of nuclear safety regulation as a separate organization under the MOE's jurisdiction. The proposed legal status was an Article 8 external bureau of the MOE, but with strengthened independence based on granting the agency staff and budget rights. Officially, responsibility for nuclear regulation would reside with the Minister of the Environment, but the competencies granted by the bill were, in principle, entrusted to the head of the nuclear regulation department. The members of the attached commission for nuclear safety were supposed to be approved by the Parliament. Apart from stipulations about the organizational setup, the draft bill contained two safety regulations, namely a forty-year limit for the lifespan of reactors and the introduction of a backfit system[4] (Cabinet Secretariat 2012b, 1–6). The safety regulations proposed by the DPJ clearly deviated from the former NISA-centered safety governance practices. Instead of licenses being practically valid for sixty years, a forty-year limit means that a second safety check and operating license are necessary in order to keep a reactor in operation. In addition, a backfit system gives a regulating agency leverage over the industry via the right to revoke a license should the utility not comply with requests for safety updates. In the draft bill, Hosono's task force included most of the recommendations from the advisory committee and the international workshop, apart from the legal status as an Article 3 independent commission organization.

A Triparty Compromise between DPJ, LDP, and Kōmeitō

In order to pass the reform bill, the Noda government depended on opposition votes, because it held a majority in the Lower House, but not in the Upper House of the Parliament, a state called "twisted Diet" (*nejirekokka*). In the Upper House, the LDP as the biggest opposition party together with its coalition partner, the Kōmeitō, held the majority of seats. It was the government's goal to pass a bill in March 2012 and to establish the new regulatory body in April to replace both NISA and the NSC. Once the DPJ submitted its draft bill to the Diet, voices within the governing party and the opposition called for the reform to await the final report of the Nuclear Accident Independent Investigation Committee, expected in June (Asahi Shimbun 2012c, 1; Hosono and Torigoe 2012, 233–34). In the weeks to come, the LDP and Kōmeitō opposition demonstrated its power in the Diet by refusing to deliberate the draft bill submitted by the DPJ. Instead, the LDP set up its own Study Group on Nuclear Regulation Reform (Study Team, *Genshiryoku-kisei-soshiki-nikansuru-purojekuto-chīmu*), chaired by Shiozaki Yasuhisa, in order to prepare a counterproposal. With LDP and Kōmeitō refusing to support the DPJ draft, deliberations in the Parliament came to a halt.

In the aftermath of the Fukushima nuclear accident, four Japanese committees to investigate the accident were established[5] and one international one. A major question for these investigations was whether the nuclear accident was in fact a "natural disaster" (*tensai*) or a "man-made disaster" (*jinsai*). A report published in February 2012 came from a private commission initiated by Funabashi Yōichi, former editor-in-chief of the *Asahi Shimbun*. The Funabashi report concluded that the Fukushima nuclear accident was a man-made disaster caused by a lack of preparations on the part of TEPCO and the nuclear safety administration. In contrast, TEPCO's June 2012 interim report stated that the Fukushima nuclear accident was the result of a tsunami of an unforeseeable size and, hence, a natural disaster. The July 2012 Hatamura report by a commission set up by the Kan Cabinet found a lack of provisions for a severe accident to be the cause. It also noted the existence of a safety myth prior to the accident, a situation in which both nuclear operators and the government thought a severe accident to be impossible in Japan. It clearly identified human failure as the reason behind the accident. The most authoritative of the four investigations, the Nuclear Accident Independent Investigation Committee (NAIIC) initiated by the Diet of Japan and led by Kurokawa Kiyoshi, also regarded the nuclear accident as man-made. The IAEA International Fact Finding Expert Mission from May until June 2011 had also concluded that the accident was due to human failure to implement suffi- cient safety provisions, particularly those aimed at natural hazards (IAEA 2011c). Hence, all reports but the interim one by TEPCO concluded the Fukushima nu- clear accident to be a man-made disaster that could have been prevented.

The NAIIC was the first-ever independent parliamentary investigation com- mittee in Japan's postwar history and influenced how the public and policymak- ers perceived the accident and its cause. The NAIIC's mandate granted it authority to request any necessary documents via the right to request the legislature to as- sist in the investigation. The interviews and hearings with almost twelve hundred people were held in the Parliament, meaning that they were public. Bit by bit, the findings that were later summarized in the July 2012 Kurokawa report (NAIIC 2012) came to light in public hearings in late 2011 and in the first half of 2012, right around the time the DPJ and the LDP/Kōmeitō opposition were arguing over reform details. For example, the former head of Japan's Nuclear Safety Com- mittee, Madarame Haruki, stated in a February hearing that "Japanese officials had succumbed to a blind belief in the country's technical prowess and failed to thoroughly assess the risks of building nuclear reactors in an earthquake-prone country." He partly blamed himself while pointing out an overall "culture of complacency" in Japan's nuclear safety governance (New York Times 2012). The Kurokawa report pointed to a lack of accident preparation, for which it directly blamed the two regulating agencies—NISA and the NSC—as well as TEPCO as the operator of the Fukushima Daiichi nuclear power plant. In its conclusion, the NAIIC found "collusion between the government, the regulators and TEPCO, and the lack of governance by said parties" (NAIIC 2012, 16). The Kurokawa re- port omitted any mention of the LDP but essentially condemned a safety gover- nance system created under many years of successive LDP governments.[6]

In February, the head of the NAIIC, Kurokawa, became involved in the de- bate about whether the new regulating agency should be an Article 3 or an Ar- ticle 8 organization. He issued a statement in which he criticized the DPJ proposal for weakening the future regulator's independence by basing its establishment on Article 8. As an NAIIC hearing on February 27, Richard A. Meserve, former chairman of the NRC, stated that the NRC and the electric utilities were in charge of handling all accident-related emergency response measures in the United States (Japan Times 2012). Since Hosono justified his preference for an Article 8 external bureau with the need for ministers and the government to be involved in emergency response measures, the statement by Meserve was a clear call for making Japan's new safety agency as independent as its US counterpart as an Article 3 independent commission organization. Thus, within the debate about the legal status of the new regulatory agency, widely recognized Japanese and international experts spoke out in favor of an Article 3 independent regulatory commission.

Throughout the reform process, public opinion on nuclear power remained skeptical. The *Asahi Shimbun* asked the public if nuclear power plants should be restarted. In March 2012, only 27 percent favored restarts, while 57 percent opposed bringing reactors back in operation, and 16 percent were unsure (Shibata and Tomokiyo 2014, 58). In January 2011 and January 2013, the Japan Atomic Energy Society conducted a public opinion poll asking how people felt about nuclear electricity generation. In those two years, the share of those stating "very anxious" rose from 11 to 32 percent and those feeling "rather anxious" went from 39 to 40 percent (Shibata and Tomokiyo 2014, 50). A majority of 72 percent expressed unease about nuclear electricity generation and 57 percent opposed moves to bring nuclear reactors back in operation.

Meanwhile, the antinuclear protests that had subsided during the winter of 2011–2012 flared up again, even larger than before. Beginning in March 2012, people gathered outside the prime minister's residence, called Kantei, every Friday. The number of participants grew during the following months and reached a peak after the government's announcement that it would let Units 3 and 4 at the Ōi power plant resume operations to ensure that peak energy demand during the hot summer months could be met. The two units had been deemed safe by the same safety regulators that had declared the Fukushima nuclear power plant safe just before the March 2011 nuclear accident. In response, on July 29 an estimated two hundred thousand people protested in front of the Kantei against the restart of the Ōi nuclear power plant, the use of nuclear power overall, TEPCO, and the government's accident response (Hasegawa 2014). The sheer number of participants was unprecedented and apparently unexpected, as the following account highlights:

> The protest was called in opposition to the restart of the Ōi nuclear power plant located in Fukui Prefecture on the Japan Sea coast that was scheduled for the following day. Although protest organizers attempted to obey police directions and restrict participants to the narrow footpath leading up to the Kantei, the sheer number of people attending the demonstrations overwhelmed both police and organizers. The crowd overflowing the temporary barriers that had been erected by the police and spilled out onto the road, eventually blocking all traffic. (Brown 2018, 148)

Due to the delay in the reform process, a vacuum emerged in Japan's nuclear administration. In line with the DPJ plan to abolish the NSC in April 2012, three of the five NSC board members resigned in April, leaving it dysfunctional. As a result, there was no procedure in place to restart nuclear power plants that had been shut down for maintenance and for safety assessments as part of the stress test. The results of the stress test needed to be approved by both NISA and the NSC in order to bring a reactor back into operation. Before the NSC practically dissolved, the Ōi power plant was the only one approved for a restart. The lack of a procedure for bringing nuclear reactors back online, at a time when there were fears that peak energy demand during the hot summer months could not be met without nuclear power plants, added urgency to the task of setting up a new nuclear regulator, especially considering the large public demonstrations against restarting the Ōi power plant.

Furthermore, looming elections framed the legislative process. The Noda government struck a deal with the opposition parties, the LDP and Kōmeitō, to cooperate in the passing of a consumption tax bill in return for a promise to dissolve the Diet (Zakowski 2015, 174–75), resulting in elections in December 2012. The LDP had developed two tactics to be an effective opposition party at the time: "criticizing individual ministers" as well as "opposing DPJ policies, forcing amendments, but passing important bills so as not to appear obstructionist to the public" (Endo, Pekkanen, and Reed 2013, 57). As the DPJ had to find a compromise between its reform bill and the one put forward by both the LDP and Kōmeitō, both opposition tactics were clearly visible in the opposition's counterproposal for a new nuclear safety agency, discussed below.

Table 1 provides an overview of four main actors' framings in the nuclear safety reform process. Diagnostic framing refers to an actor's definition of the problem and the inherent attribution of blame (Stone 1989) it entails, a crucial element of the narratives different actors seek to construct in the aftermath of a focusing event (Birkland 2007; Blyth 2002). An actor's understanding of the problem shapes the solution presented, which is known as prognostic framing. Last but not least, motivational framing expresses why an actor is seeking certain actions.

Beginning with diagnostic framing, the DPJ on the one hand and the LDP and Kōmeitō opposition on the other hand disagreed on the role of the prime minister in crisis management. Hosono insisted on giving the prime minister a leading role in emergency response management, the main lesson he drew from ANRE's and MEXT's slow response after 3.11 (Hosono and Torigoe 2012, 225–26). In contrast, Shiozaki contended, "It caused a great deal of confusion when an amateur prime minister with his superficial knowledge intervened in resolving the accident on scene. We have to exclude the 'Naoto Kan risk'" (Asahi Shimbun 2012d, 2). He attributed the distrust on the part of Japanese citizens and

TABLE 1 Overview of framings by different agents of change

ACTOR	DIAGNOSTIC FRAMING	PROGNOSTIC FRAMING (REFORM PROPOSAL)	MOTIVATIONAL FRAMING
DPJ Draft Bill	—insufficient crisis management competencies in the nuclear safety administration —a regulator unable to prevent a large-scale accident	—a unified regulating agency based on the separation of promotion and regulation (Art. 8 plus staff and budget right) —forty-year limit —backfit system —no-return rule (for managerial staff)	—regain domestic and international trust —prevent another large-scale accident
LDP draft bill and framing	—insufficient crisis management competencies by the Kan government —a regulator unable to prevent a large-scale accident	—independent regulating commission following international standards (Art. 3) —no-return rule (applying to all staff, but also a five-year grace period)	—regain domestic and international trust —enable the continued use of nuclear power —prevent another large-scale accident
Advisory Committee	—overconfidence in extant licensing and safety measures —insufficient crisis management competencies	—independent regulating commission (Art. 3) —transparency —better crisis management —no-return rule	—regain domestic and international trust —prevent another large-scale accident
International Workshop	—overconfidence in extant licensing and safety measures —insufficient crisis management competencies	—independent regulating commission following international standards (Art. 3)	—regain domestic and international trust —prevent another large-scale accident

unease in the international community not only to the accident itself but also to a great extent to the way it was handled under Prime Minister Kan's leadership (Shiozaki 2011b, 3–4). In the emergency management blame game after 3.11, the DPJ cited administrative failures, while the LDP blamed the DPJ government and Prime Minister Kan in particular.

In contrast to disagreements over the origin of crisis management problems, all agents of change eventually agreed on who was to blame for the nuclear accident itself. This was initially contested, with some investigation reports portraying it as a man-made accident, and hence foreseeable, while others insisted that it was a natural disaster, and hence unforeseeable. TEPCO painted the accident as an unforeseeable event in its interim report, which directed blame away from itself. As more investigations, and most notably that of the NAIIC, unearthed the safety governance flaws that had led to the accident, the blame game turned on TEPCO. The power company accepted the blame when its final investigation report also admitted to a man-made accident. As Samuels put it: "If Kan became the whipping boy in the discourse on leadership, TEPCO became the consensus villain in the one on risk and vulnerability" (Samuels 2013, 105). Furthermore, the different accident investigation reports pointed out flaws in the existing nuclear safety governance organizations, NISA and the NSC. While there was disagreement over the source of crisis management problems, the DPJ, the LDP, Kōmeitō, and also the participants in the advisory committee and the international workshop agreed that the accident itself was to blame on the regulator and the nuclear industry.

The blame game after 3.11 required the LDP to walk a fine line. After all, the NAIIC had not only pointed to TEPCO but also to regulators and the government. The government the NAIIC report accused of collusion at the expense of public safety was a series of LDP governments, which had previously ignored the IAEA's recommendations for a more independent safety agency and stricter regulation. Opting for an "offence is the best defense" strategy, the LDP diverted blame for the accident by drawing attention to the flawed accident response by the DPJ government. This resonated with the widely held opinion that Prime Minister Kan had not handled the accident aftermath well (Samuels 2013; Lukner and Sakaki 2011). This also aligned with the LDP's opposition tactic of criticizing the DPJ and its high-profile members (Endo, Pekkanen, and Reed 2013, 57), for example with a number of no-confidence motions by the LDP against Prime Minister Kan soon after the accident and other DPJ government members throughout 2012. This LDP strategy was not limited to blaming the DPJ, but also targeted TEPCO as public sentiment turned against the company.

Rooted in substantially different problem understandings, reform proposals by the government and the opposition differed. While the DPJ draft sought to establish an Article 8 organization in order for the prime minister to be in charge of emergency response measures (Asahi Shimbun 2012b, 3), the LDP counterproposal warned that the prime minister should not be allowed to be in charge of emergency response to preempt the "Kan risk," clearly attacking Kan's crisis mismanagement after 3.11. Similarly, the counterproposal criticized the idea of an Article 8 organization and, instead, pushed for a more independent regulatory body based on Article 3 of the National Government Organization Act as a way to conform to international standards. Both the DPJ draft bill and the LDP and Kōmeitō counterproposal stressed a unification of all functions related to nuclear safety under the umbrella of the new agency and contained a no-return rule for regulators in order to limit the kind of staff rotation that ailed NISA. The opposition's no-return rule, however, was to apply to all staff of the new agency, not only to managerial positions, but it also called for a five-year suspension of the new rule to give people a career choice (LDP 2012, 1–2). Notably absent were any stipulations regarding improved safety regulations, such as a backfit system and a forty-year limit on the reactor lifespan. This may seem to be a detail, but it was exactly those provisions that later enabled the NRA to institute top-down safety regulations. Hence, the LDP and Kōmeitō counterproposal criticized Prime Minister Kan and called for greater institutional independence, but it omitted any safety regulations that would give the new regulator leverage over the nuclear industry.

The debate about the appropriate legal status for the new agency was linked to the question of how closely Japan should follow international standards. Hosono contended that Japan should establish a Japanese-style organization, rather than simply follow the example of the US NRC, also as a way to allow for the prime minister to be involved in crisis management (Hosono and Torigoe 2012, 225–26). In contrast, Shiozaki pushed for a formally independent Article 3 organization in order to conform to international standards and to ensure that the prime minister would not be involved in times of crisis (Asahi Shimbun 2012d, 2). To address the problem, Shiozaki insisted on adhering to recommendations made by the IAEA as an international authority on nuclear safety regulation and on emulating regulatory bodies abroad, most importantly from the United States, France, and the United Kingdom (LDP Bulletin 2012). The idea of a formally independent Article 3 organization was supported by those in the government's advisory committee and international workshop participants. While Hosono advocated for a Japanese-style institution inspired by international norms of regulatory independence, Shiozaki and other domestic and international policy entrepreneurs argued that the best way to regain domestic and international trust was to make the agency legally independent in line with international standards.

LDP support for the creation of an independent safety agency stood in contrast to the party's history as a pillar of the nuclear village and champion of safety reforms that resulted in mere window dressing. Such a regulatory body with a higher degree of independence and far-reaching competencies was initially viewed critically within the LDP with its numerous supporters of nuclear power, many of whom cautioned against the establishment of an independent nuclear regulation organization (Asahi Shimbun 2012b, 3). Eventually, however, the idea promoted by Shiozaki was adopted. For those in the LDP who feared that the MOE—traditionally more critical of nuclear power than METI—might curtail nuclear power too much (interview with Japanese Energy Industry Expert, 2015), this plan presented a way to avoid the integration of nuclear safety into the MOE by attaching the regulator only loosely. Pushing for more independence also tied in nicely with the aims of regaining domestic and international trust, diverting blame for the accident away from the LDP, and lambasting the incumbent government in the run-up to elections.

Despite the differences in problem understanding and the solutions presented, one aspect the different agents of change agreed on was the motivational framing, meaning the reasons why actions should be taken. The proposals and statements by the DPJ government, LDP reform leader Shiozaki, the advisory committee, and the international workshop all stressed the need to regain both domestic and international trust in Japan's nuclear safety governance. Clearly, nuclear safety reforms in Japan had two key audiences, one domestic and one international.

Once both draft bills were submitted, there was about one month left in the parliamentary session, putting pressure on the government and the opposition to swiftly pass reform legislation. Had a decision about the NRA been pushed into the next Diet session, it would have left Japan without a nuclear administration at a time when the damaged Fukushima Daiichi power plant needed supervision and all but two commercial reactors were shut down for a safety review. In three-party consultations that included the DPJ, the LDP, and Kōmeitō, the debate revolved around the most appropriate legal status for the new regulatory agency. In early June, the DPJ consented to the LDP and Kōmeitō's demand of making it an Article 3 independent regulatory commission (Asahi Shimbun 2012e, 27), but the issue of which role the prime minister should play in a time of crisis remained contentious. The compromise found under time pressure from the nearing end of the parliamentary session was to limit the role of the prime minister to operational aspects of emergency response holding political responsibility but based on expert opinions and assessments by the new regulating body (Asahi Shimbun 2012f, 1).

With the legal status and role of the prime minister settled, the last remaining issue was the selection of members for the regulatory agency's board of scientists. As part of the three-party consultations, the DPJ, the LDP, and Kōmeitō agreed that commissioners should disclose the amount of contributions and research grants that they received from the electricity industry (Asahi Shimbun 2012f, 1). Meanwhile disagreement remained over the amount of transparency in the process, that is, how much information should be made available to the public, with the DPJ asking for greater transparency than the LDP was willing to commit to. As a result, selection criteria remained vague.

One June 20, one day before the end of the parliamentary session, the DPJ, the LDP, and Kōmeitō passed the NRA Act, which stipulates that the new regulatory body shall be established as an Article 3 independent regulatory commission as requested by the LDP. Furthermore, in case of an emergency, the prime minister is to be in charge of operational aspects of crisis response while the new agency is to handle technical aspects. What the law does not contain are any of the DPJ's propositions regarding new safety measures, such as a forty-year limit to the reactor lifespan or a backfit system. As the DPJ conceded to most demands by the LDP except for the issue of which role the prime minister should play in times of crisis, the NRA Law clearly bears the hallmarks of the LDP counterproposal with its focus on institutional independence while omitting new safety standards.

Concretely, the NRA Act amends the Atomic Energy Basic Act (Act No. 186 of 1955), the Act on the Regulation of Nuclear Source Material, Nuclear Fuel Material and Reactors (Act No. 166 of 1957, the "Reactor Regulation Act"), and the Act on Special Measures Concerning Nuclear Emergency Preparedness (Act No. 156 of 1999, "Nuclear Emergency Act"). Amending the Atomic Energy Basic Act was necessary to take nuclear safety regulation functions out of METI and to effectively integrate the nuclear safety functions of METI's NISA, the NSC, and JNES into one new organization affiliated with the Ministry of Environment. Changes in the Reactor Regulation Act resulted in transferring nuclear safety regulation competences to the NRA's jurisdiction. While ANRE and the AEC remained in charge of devising nuclear policy, the NRA became solely responsible for nuclear safety regulation. The Nuclear Emergency Act, together with the Basic Act on Disaster Control Measures, forms the legal framework for emergency response measures in case of a nuclear accident. Amending it put the NRA in charge of technical aspects of emergency response measures. With the NRA Act, a legal framework was put in place for an independent unified regulator outside of METI.

In parallel to safety regulation reforms, debates about Japan's future energy policy, especially the role of nuclear power, took place. Unlike for Kan, energy policy was not a matter close to Noda's heart, who was more focused on other policy reforms, such as the consumption tax increase. Noda entrusted energy policy deliberations to METI Minister Edano Yukio and State Minister for Economic and Fiscal Policy (as well as Science and Technology) Policy Furukawa Motohisa—both of whom favored a phaseout. Rather than take matters into his own hands,

Noda followed the lead of Furukawa and Edano on energy issues. Noda appointed Edano as head of the Energy and Environment Policy Committee (EEC). The EEC deviated from the usual practices of energy policy deliberation committees under METI's guidance. The committee of about twenty-four people started out with only two or three critics of nuclear power. Edano soon raised the ratio of critics to about one third of the members. In addition, he asked government officials to hold back and not push deliberations toward a certain result, resulting in an unusually open and frank discussion about the pros and cons of nuclear power with many outside observers and critics of nuclear power (interview with DPJ Government Adviser, 2015). Such discussions set the EEC apart from other policy councils, whose members are often selected by public officials to support a predetermined goal (Schwartz 2010).

In addition to discussing future energy policy under cabinet leadership, the DPJ gave the AEC a bigger role in the debate. Throughout the history of Japan's nuclear power program, METI had chipped away at the AEC's power in the energy policymaking process. The DPJ, however, actively engaged the AEC in its attempt to reformulate Japan's nuclear energy policy from scratch. As a former AEC member recollected, the AEC discussed different scenarios ranging from a 0 percent to a 50 percent mix of nuclear power in 2030, and some members were eager to think about long-term goals beyond 2030. This was met with opposition by METI, which claimed responsibility for all energy policymaking for itself and refused to talk about time periods after 2030. METI successfully limited the time frame of discussions up to 2030, but to its dislike, it failed to claim sole energy policymaking responsibility. During this struggle, parallel debates took place in the AEC and METI as opposed to the consecutive debates before 2011, where the AEC often had to follow METI's lead (interview with Former AEC Official, 2015). Adopting METI's line to limit discussions to 2030, the AEC developed three scenarios for the EEC.

In June 2012, the EEC published these three scenarios for Japan's future nuclear power policy—a 0 percent, 15 percent, and 20–25 percent share in Japan's 2030 electricity mix—and put them up for public discussion. The DPJ government sought to include citizens in energy policy deliberations, for which it held public hearings and conducted deliberative polling.[7] As a first-time experiment in Japanese policymaking, these discussions experienced similar problems as previous local hearings about nuclear power plants had. One time, a speaker in favor of nuclear power turned out to be a utility employee who claimed to speak on his own behalf but sounded as if the electric utility had scripted his statement. This was reminiscent of pre-3.11 practices of influencing public hearings. Nevertheless, the result of citizen participation was a clear majority in favor of a 0 percent scenario, not the 15 percent scenario the DPJ had expected and hoped for (interview with Former AEC Official, 2015; interview with DPJ Government Adviser, 2015).

In parallel to the deliberative polling exercise, large daily newspapers conducted public opinion polls to gauge which option garnered the most public support. Polls by the center-left *Asahi Shimbun* showed a clear majority favoring the 0 percent scenario from the beginning. The results published by the conservative pronuclear *Yomiuri Shimbun* showed a move away from almost half the respondents favoring the 15 percent scenario to a tie between 0 percent and 15 percent. The 20–25 percent scenario received the least support in both polls (Shibata and Tomokiyo 2014, 76). Just as in the public hearings and deliberative polling, public opinion polls showed major support for the scenario of 0 percent nuclear power by 2030.

In mid-September, the EEC approved the "Innovative Strategy for Energy and the Environment" (*Kakushinteki-enerugī-kankyō-senryaku*) report, which included a commitment to achieving zero nuclear power in the 2030s. The report was not translated into a Cabinet Decision on a nuclear phaseout. Rather, the Noda Cabinet adopted the vague resolution "to reduce dependence as much as possible" on nuclear power that could be interpreted as any of the three scenarios previously discussed. This decision to backtrack from a concrete phaseout plan is widely attributed to severe criticism from the Japanese business lobby (Homma and Akimoto 2013; Kingston 2014; Edahiro 2012; interview with DPJ Government Adviser, 2015). Looking at the overall picture, the failure to adopt a Cabinet Decision on a nuclear phaseout was the result of the public pushing for a complete phaseout that the DPJ at large and the Noda Cabinet itself did not stand behind and business circles vehemently opposed.

By late summer of 2012, the debate about Japan's nuclear energy policy and regulation had culminated in two decisions. First, the DPJ government resolved to reduce Japan's dependence on nuclear power without making a binding decision by casting it into a Cabinet Decision. Second, the DPJ, with the backing of the LDP and Kōmeitō passed the NRA Act in order to improve nuclear safety governance.

Preparing the Creation of the NRA

Final arrangements for the new agency were made by the preparation office at the Cabinet Secretariat, which had also handled the bill-drafting process and was headed by the MOE official Morimoto from the beginning. They drew up concrete plans for the organizational structure of the NRA, the emergency countermeasures system, human resources distribution, and so forth. Split into a number of working groups, it consisted of about twenty-five to thirty people, including those who were previously drafting the DPJ reform bill (interview with NRA Official, 2018). Those in the preparation office expected much scrutiny once the Diet convened again in October 2012, and thus felt it was important to arrange everything beforehand (interview with Preparation Office Member, 2020). Final arrangements also included selecting experts for the NRA board.

The NRA as an organization consists of two bodies, an elected board called the NRA Commission and a supporting Secretariat. The role of the board is specified by the NRA Act to consist of five members, called commissioners, one of whom acts as the NRA chairman. To ensure board members' independence from the remainder of the administration, they are not appointed by ministry officials. Rather, they are appointed by the prime minister and need Diet approval. In cases where Diet approval cannot be obtained, the prime minister can preliminarily appoint board members but must later obtain ex post facto consent by the Diet (Article 7). Eligible candidates cannot include persons with close linkages to the nuclear industry (Article 7). Board members are prohibited from holding other positions while on the NRA board (Article 11). These provisions clearly set them apart from members in other commissions in Japan's government and administration, who are appointed by the responsible ministry and hold full-time positions in a related field, whereas being an NRA board member is a full-time position in itself. The board is the highest decision-making organ within the NRA and decides any aspect not specified in the NRA Act (Article 28). Given their far-reaching competencies to shape the new regulatory agency and to make regulatory decisions, choosing board members was a crucial piece of the puzzle.

For the appointment of board members, Hosono presented guidelines that specified Article 7, which remained vague after disputes between the DPJ and the LDP over the appropriate amount of transparency. The DPJ selection guidelines aimed to "distinguish clearly between those with ties to the electric utilities and those without. As such, no members of the nuclear village will be chosen" (Asahi Shimbun 2012g, 4). Concretely, they excluded anyone who had worked in the nuclear industry or in a related association in the previous three years or received more than ¥500,000 a year in salary from either a company in the nuclear industry or a related association. The guidelines also required the disclosure of information about receiving grants or sponsorships for research with ties to the nuclear industry as well as about students taking up a position in the nuclear industry (Asahi Shimbun 2012h, 4). DPJ selection criteria were designed to ensure a certain distance between top regulators and the industry they oversee.

To come up with candidates for Noda to make a decision on, Hosono gathered a small group of about five people (interview with Former Administrative Vice Minister, 2015). The five candidates Noda presented to the Diet were Fuketa Toyoshi, Nakamura Kayoko, Ōshima Kenzō, Shimazaki Kunihiko, and Tanaka

Shun'ichi. The choice of Tanaka Shun'ichi was particularly controversial. Tanaka was criticized by Diet members from different parties as a "member of the nuclear village" because he had worked for the Japan Atomic Energy Agency JAEA) for many years (Asahi Shimbun 2012k). When Noda was unable to attain parliamentary approval in early September, he refused to change nominations, pointing out that they presented the best option possible (Asahi Shimbun 2012j). Hosono wanted Tanaka Shun'ichi to become NRA Chairman, as he later explained in his book. As decontamination adviser to the MOE, Tanaka had shown that he was willing to take responsibility—more than other experts—without fear of a possible backlash (Hosono 2013, 147–48). This willingness to not only comment from a distance, but also to become involved in an official function, set him apart from many other experts at the time (interview with Preparation Office Member, 2020). Also, there was a limited pool of candidates. Even people from within the Cabinet Office admitted that it was difficult to find people who fit the DPJ selection criteria by being experts in the field without being too close to the industry (Asahi Shimbun 2012i). In the end, Noda preliminarily appointed board members without parliamentary (i.e., LDP and Kōmeitō) approval.

Prior to their preliminary appointment, the designated candidates already worked as cabinet advisers in the preparation office. The preparation office had prepared draft guidelines for the NRA's day-to-day operations and discussed them with the designated candidates. In little more than a week, the designated candidates and the preparation office finalized operational guidelines for the NRA (interview with NRA Official, 2018). At a preparatory meeting on September 12, held at the MOE, Chairman-designate Tanaka (still labeled a cabinet adviser in the meeting minutes) presented transparency and neutrality guidelines, which suggested making a transparency requirement the first supplementary provision to the NRA Law, which the participants agreed to. Opening the preparatory meeting, Tanaka stated that "the regulatory agency should attribute utmost importance to transparency and openness" (MOE 2012, 7). Implementing this policy from the beginning, the press was invited to observe the meeting, and a video recording was published afterwards. Hence, designated board members and the preparation office at the Cabinet Secretariat moved swiftly to fill in the details left unspecified by the NRA Act with guidelines for transparency and neutrality.

Before the elections in December 2012, the DPJ government and the NRA board amended the NRA Act with supplementary provisions. As an Article 3 independent commission organization, the NRA had the right to suggest amendments to the law that pertained to its organizational practices. In practice, the NRA handed them over to Minister of the Environment Hosono to introduce them into the Parliament. The first supplementary provision introduced pertained to transparency as a means of "guaranteeing the public's right to know" (Article 25 of the NRA Act). Another amendment concerned staggered terms of office for the first board members, with two appointed for two years, another two for three years, and only the chairman serving a full five-year term. This avoided all board members being replaced at the same time to ensure continuity. With another supplementary provision, the NRA and the Noda government tasked the government with abolishing JNES so that the NRA could take over its functions. Since JNES was in charge of monitoring safety at nuclear power plants, integrating it into the NRA was equivalent to the new safety regulator acquiring monitoring competencies. As early as November 2012, the Noda government called a Vice Administrative Minister's Meeting to discuss the details (Cabinet Secretariat 2012a). Later in February 2012, the NRA board adopted the forty-year rule and the backfit system, both of which had not made it into the NRA Act due to LDP opposition. These additions to the NRA Act and early board decisions ensured the longevity of the first board's understanding of independence and improved safety and became cornerstones for safety governance by the NRA (discussed in detail in chapter 3).

The LDP won the December 2012 elections in a landslide victory. Technically, it had a chance to replace the whole NRA board, because it still required Diet approval. Instead, the newly elected, openly pronuclear Abe government ex post facto approved the board in the Diet on February 15, 2013. While replacing the first board was technically easy for the Abe government with its Diet majority, it would have undermined the credibility of its commitment to independent and trustworthy nuclear safety governance. Starting with his New Year's Address (Abe Shinzo 2013), whenever Prime Minister Abe spoke of nuclear reactor restarts, he stressed that this was preconditioned on the NRA as an independent regulatory agency certifying reactors' safety. In the face of a critical domestic and international audience and the LDP's strong push for an independent agency, openly undermining the NRA's independence became a taboo for the LDP government in order to remain credible in its commitment to nuclear safety.

Change in Safety Governance, Continuity in Nuclear Energy Policy

This chapter aimed to understand why 3.11 induced fundamental change in Japan's nuclear safety governance institutions but no nuclear phaseout policy, and what this reveals about mechanisms for change in the aftermath of a shock event. To do so, it traced *which* decisions were influential, *who* was behind them and *why* it was adopted rather than an alternative one.

In each case, there were three influential decisions. In the summer of 2011, the DPJ cabinet under Prime Minister Kan adopted a Cabinet Decision for a unified safety regulator with stronger in-house expertise and regulatory competencies to be located outside of METI. This was an important first step that set the reform trajectory. Subsequently, both the DPJ government and the LDP/Kōmeitō opposition produced draft bills for a new regulator, and they found a compromise in the summer of 2012. In the process, the LDP pushed through Article 3 as a strong legal foundation for a legally independent regulatory organization but at the same time forced a removal of concrete governance rules, such as transparency and new safety regulation, which would limit utilities' power. Once the NRA Act was passed, the DPJ government moved quickly to finalize reform details before upcoming elections. Working closely with designated NRA board members and senior officials, they used the freedom granted by Article 3 to introduce the very elements the LDP had opposed into the guidelines for the new agency and even added some, such as transparency, to the NRA Act. By the time the LDP returned to government, the new regulatory agency was in full operation based on guidelines for independent, neutral, and transparent safety governance and working on expanding its regulatory and monitoring competencies. The three influential decisions in the reform process were 1) to locate it outside of METI, 2) to make it an Article 3 organization, and 3) to use Article 3 for the creation of a transparent safety regulator with binding safety standards and monitoring competencies.

Also in the summer of 2011, the DPJ government resolved to identify ways for Japan to phase out nuclear power. The phaseout debate took place under the Cabinet Office in parallel to the safety administration reform, but there was neither a Cabinet Decision nor a draft bill introduced into the Diet. Ultimately, it produced a vague DJP government decision to reduce Japan's dependence on nuclear power. In contrast to safety administration reforms, energy policy deliberations fell back into the hands of the pronuclear LDP and METI. With the 2014 Strategic Energy Plan they translated the idea of a reduction into a policy goal of a 20–22 percent share of nuclear power in the 2030 electricity mix. This is less than half of what the 2010 Strategic Energy Plan aimed for, which was at least 50 percent (METI 2010), and less than the 29 percent share that nuclear power made up just before the nuclear accident, but far from the phaseout the public supported in 2012. A difference that jumps out was the finalization of the nuclear safety administration reform process before public mobilization ceased and the DPJ was voted out of government, which marked the end of a window of opportunity for radical change. The establishment of the new safety agency was concluded before the window of opportunity closed, but the bigger energy policy question was taken over once again by actors vested in nuclear power.

Ideas as a Change Mechanism

One explanation for why reform initiatives following a shock event produce different outcomes lies in the existence of an alternative idea around which the different agents of change consolidate. Looking at the nuclear safety governance reforms versus the idea of a nuclear phaseout, a readily available alternative existed solely for the safety governance reform. The IAEA had given recommendations to Japan in 2007 and, based on these, a study group had produced suggestions for a unified and more independent regulatory agency. Ignored at the time, they were available to agents of change to model their reform proposal on. Furthermore, a clear norm for independent nuclear safety regulation existed at the international level, which mattered as a reform model as Japanese decisionmakers looked to the international community as one key audience during reforms. In contrast, there was no readily available alternative energy policy strategy that dispensed with nuclear power and no single international model to follow. Rather, it was a question fraught with uncertainties about how quickly Japan would be able to replace nuclear power with renewable energy sources, the impact it would have on energy costs and greenhouse gas emissions, and the implications this would have for Japan's energy security. Thus, available alternative ideas were clearly available with safety governance reforms but not nuclear power overall.

Over the course of 2011 and 2012, more agents of change came to support the idea of a unified and independent nuclear safety agency as a means to regain domestic and international trust. These included not only the DPJ government but also the Japanese public, along with Japanese and international experts. The LDP loudly advocated for even stronger legal independence than the DPJ in order to divert blame for the accident away from itself and to lambaste the DPJ government in the run-up to elections. METI and powerful industry associations showed few signs of opposition to the proposed administrative reform, especially compared to their reaction to the phaseout idea. In a twist, the DPJ government and designated NRA board members and senior officials introduced transparency and concrete safety measures the LDP had opposed, adding substance to the independence stated in the legal framework. In sum, all three categories of agents of change—political entrepreneurs (DPJ and LDP), policy entrepreneurs (Japanese and international experts), and outside actors (Japanese public and international actors)—supported the idea of independent nuclear safety governance. With a concerted push by domestic and international agents of change for the safety administration reform the DPJ had initiated, power constellations were favorable for that reform idea.

There was an intense debate about a nuclear phaseout and the changes to Japan's energy policy required to achieve it. The DPJ government, which was united behind nuclear safety governance reforms, held diverging views on the future of nuclear power. This was unsurprising considering the daunting nature of the task, which was nothing less than to reconsider half a century of nuclear power promotion as part of Japan's energy policy. Furthermore, Japan's nuclear safety administration had been reformed before without actually producing much change, making it a less daunting endeavor than radically changing energy policy direction, as would have been the case with a nuclear phaseout. While the DPJ, especially in the person of Prime Minister Kan, acted as a political entrepreneur by putting the idea on the political agenda, internal divisions and nuclear industry opposition resulted in little more than a vague resolution to reduce reliance on nuclear power despite public opinion pushing for a swift phaseout. Actors historically vested in nuclear power, METI and the LDP, were able to take over energy policy discussions once the DPJ government resigned at the end of 2012. Thus, the DPJ government put a nuclear phaseout on the agenda, but agents of change failed to consolidate around the idea and Japan's nuclear triangle was ultimately able to defend the nuclear policy option.

One potentially influential outside actor, the media, played a minor role in the reform processes. If the media pick up and push an issue, they can act as agenda-setting outside actors. In Japan, however, mainstream media ignored early warnings of a nuclear meltdown by Japanese scientists (McNeill 2014). Negative coverage of the nuclear industry was largely suppressed by the advertising powerhouse Dents' before and during the March 2011 nuclear accident (Honma 2012). Hence, newspaper reporting about the accident was largely limited to presenting official government statements (Segawa 2011). Furthermore, mainstream media failed to pay significant attention to protests in front of the Kantei until the numbers reached the hundreds of thousands in mid-2012 (Brown 2018). As a result, Japanese mainstream media were not part of the pool of agents of change and not a potential source for alternative ideas.

Power as a Change Mechanism

Another explanation for policy changes in the aftermath of shock events focuses on whether agents of change have the power to push through reforms. The political power constellation in Japan was rather unusual, because the long-term governing LDP was in the opposition. At the same time it held a majority in the Upper House together with its coalition partner Kōmeitō. This "twisted Diet" forced the DPJ administration to compromise with the opposition to pass legislation. Since the LDP opposed a nuclear phaseout, it would not have been possible for the DPJ to pass legislation to that end. The alternative, adopting a Cabinet Decision, which would have forced the successor LDP government to earnestly engage with the idea, however, was also not in the power of the DPJ due to internal divisions.

In contrast to a nuclear phaseout, both the DPJ and the LDP supported nuclear safety reforms. Ahead of elections, the LDP-led opposition used reforms as a means to oppose DPJ policies while proposing alternatives to avoid being seen as obstructionist, which was its main opposition party strategy (Endo, Pekkanen, and Reed 2013). The LDP alternative was legally enshrined independence of nuclear safety regulators in line with the independence norm promoted by the IAEA as the supposedly best way to regain domestic and international trust, a major concern for the LDP and most other agents of change. The LDP and its partner Kōmeitō used their majority in the Upper House to push through Article 3 as the legal foundation for the new safety agency. The DPJ, in turn, used its power in government to allow designated NRA board members and senior officials, including one former NAIIC member, to essentially run with the independence granted by the Article 3 status and set the agency up the way they saw fit. This collaboration between the DPJ and NRA in the establishment phase in late 2012 crucially shaped nuclear safety reforms in such a way that they became political game changers.

It mattered that the LDP was the largest opposition party but not in government at the time. This becomes clear when looking at the counterfactual, that is, "what happened in the context of *what could have happened*" (Berlin 1974, 176, italics added by Capoccia and Kelemen 2007). Counterfactual analysis as a method to support a causal explanation invokes other established regularities, theories, or principles found, for example, in existing literature (Fearon 1991, 177). It may also involve a comparison with a hypothetically plausible alternative that feeds on the researcher's background knowledge (Collier 2011, 825). Thus, a counterfactual analysis helps to establish the respective historical weight of a certain decision in determining the observed outcome by assessing whether another decision would have led to a similar outcome. What reforms may have looked like under LDP leadership becomes clear when looking at previous nuclear safety governance reforms and the details of the counterproposal the LDP tabled in 2012.

Had the LDP been in government at the time, nuclear safety administration reforms would have looked different. Previous reforms integrated more and more safety functions into METI's energy agency NISA. They were, however, mere window dressing as they failed to address, and in some instances even exacerbated, METI's influence and that of the nuclear industry over safety regulation. Reversing this trend by taking nuclear safety functions out of METI became possible because the Kan Cabinet early on preempted a METI veto to such a move. Considering previous LDP-led reform trajectories and how closely LDP and METI worked together in the subsequent LDP government under Abe, such a move would have been highly unlikely. Rather, safety administration reforms by the LDP would have likely produced a safety agency similar to ANRE, the semi-independent energy policymaking agency affiliated with METI. In the counter-bill, the LDP pushed for a legally independent agency but also sought to exclude binding regulation and transparency, just as under the previous system of partial industry self-regulation with little public accountability. It should be noted here that the legal status of Article 3 alone is not a guarantee for agency independence as the "notoriously weak" Japan Fair Trade Commission (Vogel 2006) illustrates. Furthermore, the LDP had initially opposed the choice of NRA board members, who later turned out to be zealous reformers that removed the NRA from standard Japanese practices, such as officials retiring to the industry they used to oversee. Since the LDP has a history of cooperating with METI, was not pushing for transparency, a forty-year rule, or a backfit system, and opposed Tanaka as NRA Chairman, it would have most likely created a unified regulator based on Article 3, loosely affiliated with METI with less transparency and less binding safety standards. Since transparency, binding safety measures, and the repudiation of traditional practices are major factors that contribute to the NRA being a game changer in nuclear politics, an LDP government would have most likely not broken with the practices of the nuclear village but given it the façade of more independence.

The findings speak to the long-standing debate about who governs Japan, politicians or public officials. While early studies stressed the power of the bureaucracy, particularly METI, in shaping Japan's economic and energy policy (Johnson 1982, 1995; Vogel 1999), the Hashimoto reforms empowered the prime minister and the government to lead the policymaking process (Krauss and Nyblade 2005; Poguntke and Webb 2005). In 2009, the DPJ came to power with the aim to empower politicians vis-á-vis elite bureaucrats by removing them from the decision-making process, which was embodied in the slogan of "political leadership" (*seiji shudō*) (Kushida and Lipscy 2013b; Zakowski 2015). Throughout the reform process, the DPJ government relied on an instrument introduced by the Hashimoto reforms, namely that of a state minister for specific tasks. First, Kan put nuclear safety administration reforms in the hands of Hosono as State Minister for Nuclear Affairs. Hosono remained in that position under Noda while also assuming the post of Minister for the Environment. The DPJ government also excluded both the prospective reform winners and losers in the bureaucracy, namely the MOE, METI, and MEXT, from the process, allowing for continued DPJ leadership.

Of course, public officials, mainly from METI, MOE, and the police agency, worked inside the Cabinet Office to prepare a Cabinet Decision, to formulate a draft bill, and to prepare the NRA inauguration. A number of these officials went on to assume high-level positions in the new safety agency, where they asserted the NRA's independence from the remainder of the bureaucracy as well as other governance lessons learned from the March 2011 nuclear accident (discussed in chapter 3). Overall, the degree to which ministries were involved in the process can be considered very low, because they played a limited role as policy entrepreneurs, and representatives of nuclear village members were not given the chance to reassert their interests. The DPJ took a similar approach to energy policy deliberations, where it invited the AEC to play a larger role than usual, to the dismay of METI, and consulted the public on different policy options. Rather than public officials, the DPJ invited fresh faces to act as policy entrepreneurs into the process, including critics of nuclear power and outside actors such as the IAEA and representatives from foreign nuclear safety agencies. Hence, the DPJ government lived up to its "political leadership" slogan by keeping reforms close to the cabinet and by empowering different voices than before.

A Critical Juncture in Safety Governance Institutions

The findings about ideas and power reveal that they play out in connection to one another. Actors rallying around one reform idea can shift power constellations. The first example concerns organizations (collective actors) as the main actors in politics. If members of a collective actor, such as a government, are divided about the appropriate reform direction, they will hardly be able to shape it. Had all Kan Cabinet members agreed on the need for a nuclear phaseout, a Cabinet Decision to that end would have put pressure on the LDP to at least engage with the idea in the Diet. All the members of the Kan Cabinet witnessing safety administration flaws firsthand had united them behind a nuclear safety administration reform, which enabled them to determine the fundamental reform thrust with a Cabinet Decision. They decided to remove safety functions from METI's jurisdiction, which was the exact opposite of the direction previous reforms had followed. The second example concerns voting majorities in the Parliament. More agents of change coalescing around the idea of a unified and independent nuclear safety administration outside of METI put pressure on the LDP. One notable source of pressure was international distrust in the accident's aftermath and the push to adopt the international norm of independent nuclear safety governance. The LDP supporting the DPJ initiative shifted power constellations in favor of adopting reform legislation. Thus, ideas and power are connected in two ways: when all members of a collective actor unite behind one reform idea, this increases their potential to shape reforms; and more agents of change supporting an idea can incentivize a powerful actor to push for reforms it would otherwise have opposed.

Historical institutionalist studies take a long-term perspective because institutional changes take time to put down roots and to unfold their full potential. With regards to how much of a difference the "realist variant" of nuclear safety governance—with its focus on improvements in safety measures and transparency as a necessary means to reduce risk levels—would make (Samuels 2013, 115), there was apprehension of strong pronuclear interests attaining influence over nuclear safety regulation once again (Shadrina 2012; Aldrich 2014; Hymans 2015). More than a decade after 3.11, it has become clear that nuclear safety reforms created a Trojan horse in the nuclear village. Despite the intention to enable the continued use of nuclear power with improved safety governance, the NRA's stringent and transparent safety governance, in fact, has fundamentally altered the politics of nuclear policy implementation. The NRA marked the beginning of a new safety governance path for Japan because the safety governance mode it employs fundamentally changed the power distribution in favor of regulators and the public, undermining the institutions that unified and empowered the nuclear triangle.

THE NUCLEAR REGULATION AUTHORITY (NRA)

> Our fundamental mission is to protect the general public and the
> environment through rigorous and reliable regulation of nuclear
> activities. We in the NRA and its supporting secretariat shall
> perform our duties diligently in accordance with the following
> principles: 1) independent decision making, 2) effective actions,
> 3) open and transparent organization, 4) improvement and
> commitment, 5) emergency response. (NRA 2013b, 3–4)

Such reads the mission statement of the Nuclear Regulation Authority (NRA),
created in 2012 as an independent organization, affiliated with the Ministry of
Environment, and endowed with sole authority over nuclear safety in Japan. The
NRA Establishment Act (NRA Act, *Genshiryoku-kisei-i'inkai-secchi-hô*), passed
in June 2011, integrated all nuclear safety–related functions into the NRA. Until
3.11, nuclear safety functions were shared between three bodies: NISA, the Nuclear Safety Commission (NSC), and the Japan Nuclear Energy Safety Organization (JNES). The main one, NISA, was integrated into the Agency for Natural
Resources and Energy (ANRE), the energy policymaking agency within METI.
A recurring theme during the reform process was the creation of a single agency
outside of METI in order to separate the promotion and regulation of nuclear
power and to unify dispersed nuclear safety administration functions. Reforms
were motivated by a desire to prevent another accident and to regain trust in Japan's nuclear safety governance, both within the country and internationally. Ac-
cordingly, the NRA mission statement also begins by stating its aim to regain
domestic and international trust (NRA 2013b). The change that resulted from
reforms in the nuclear safety administration is depicted in figure 3.1. Note that
nuclear energy policymaking bodies are depicted with a continuous line and nuclear safety bodies with a dashed line, while host bodies are depicted in bold lines.

The NRA's predecessors, NISA and the NSC, had limited regulatory powers
vis-à-vis the nuclear industry, depended on industry expertise, lacked public accountability, and gave little voice to experts calling for stricter safety measures.
They also lacked a regulatory framework granting the competencies necessary

BEFORE

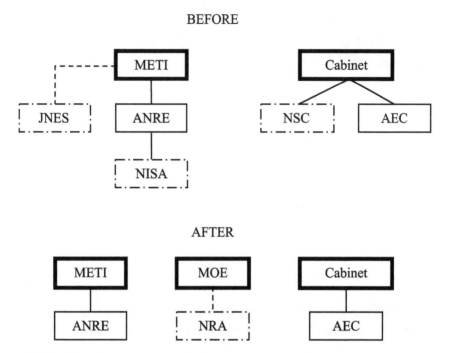

FIGURE 3.1. Nuclear safety administration before (top) and following (bottom) reforms

for top-down regulation after early attempts at state control were met with fierce opposition by the nuclear industry. A lack of in-house expertise resulted from regular staff rotation with those parts of the administration promoting nuclear power and a lack of funding for building up in-house expertise. This further enabled influence from nuclear industry firms that dispatched experts to work temporarily inside the nuclear safety administration. Despite reform attempts before 2011, Japan's nuclear safety governance institutions suffered from regulatory capture and a lack of public accountability (discussed in chapter 1).

Hurdles to Overcome

Given this history, it is unsurprising that many harbored doubts about post-3.11 reforms producing a result different from previous reforms that resulted in window dressing. Doubts about the NRA's chances of maintaining its independence were rooted in the dominance of pronuclear actors over nuclear energy policy, pressure from the nuclear village to grant safety permits swiftly, staff being taken over from the previous safety administration, and a pronuclear chairman heading the new agency (Aldrich 2014; Cotton 2014; Vivoda and Graetz 2014; Kingston 2014; Shadrina 2012). Against all odds, the NRA soon began to act like an independent regulatory agency. It was able to defend its independence and to firmly establish fundamentally different safety governance despite pressure from nuclear proponents to return to pre-3.11 business as usual.

Understanding how the NRA overcame this legacy requires a closer look at its organizational characteristics and how they enabled its resistance against political, administrative, and industry pressure. Organizational characteristics encompass both formal and de facto independence, referring to the legal framework and effective day-to-day autonomy in their regulatory activities. An organization's formal independence depends on 1) how the chairman and the management board operate, 2) what the agency's formal relationship with elected politicians is, 3) how it finances and organizes itself, and 4) what its regulatory competencies are. An agency's effective autonomy in day-to-day interactions is determined by the amount of outside influence on the budget, regulatory decisions, internal organization, and staff composition (Gilardi and Maggetti 2011). The point about staff composition reflects Wilson's (1989) notion that staff exchanges between regulator and industry and a lack of in-house expertise further the industry's ability to influence the regulatory agency. Last but not least, public accountability (Jordana, Fernández-i-Marín, and Bianculli 2018), ideally involving public participation (Beck 1992), matters for an agency's independence.

The Board's Rules of the Game

The NRA as an organization basically consists of two bodies: an elected board called the NRA Commission, the highest decision-making organ within the NRA, and a supporting secretariat. In contrast to members in other commissions in Japan's government and administration, which are usually appointed by a ministry and hold full-time positions in a related field, NRA board members hold a full-time position and require approval by the Parliament. Given their far-reaching competencies, the choice of board members was crucial.

Finding experts without close linkages to the industry for the NRA board proved a challenge. After decades of close ties between industry and academia, many experts had received research funding or a salary from plant manufacturers, plant operators, industry associations, or related foundations. Hence, there were not many who met the selection criteria presented by the DPJ government in 2012, which stipulated that candidates may not have worked in the nuclear industry or a related association in the previous three years or have received more than ¥500,000 a year from either. Furthermore, candidates were required to disclose information about grants or sponsorships for research with ties to the nuclear industry.

The first five board members were Fuketa Toyoshi[1] as commissioner for safety of nuclear facilities, Nakamura Kayoko[2] as commissioner for emergency response and protection from radiation, Ōshima Kenzō[3] as commissioner for nuclear security and safeguards, Shimazaki Kunihiko[4] as commissioner for earthquake and tsunami countermeasures, and Tanaka Shun'ichi[5] as NRA chairman. When Prime Minister Noda was unable to attain Diet approval for their appointment, he used his privilege as prime minister to preliminarily appoint them without Diet approval.

The choice of Tanaka Shun'ichi was particularly controversial. He was criticized by Diet members from different parties as a "member of the nuclear village" because he had worked for the Japan Atomic Energy Agency (JAEA, formerly Japan Atomic Energy Research Institute, JAERI[6]) for many years (Asahi Shimbun 2012k). Contrasting those views, Tanaka later described himself as a "misfit in the nuclear power realm, who has often been called a member of the nuclear village, but really had been tolerated rather than embraced by the village community" (Toyo Keizai 2018). The story of Tanaka reveals the difficulty of discerning advocates of transparency and stringent safety regulation from those who preferred to return to business as usual. More generally, it highlights the difficulties of clearly delineating the boundaries of the nuclear village, which is not monolithic and where membership can be a matter of perception. For Tanaka, the aftermath of 3.11 was not the first time he had dealt with a nuclear accident. He had been in charge of handling the aftermath of the JCO Tōkaimura critical event in 1999. Following the Fukushima Daiichi nuclear accident, he also became decontamination adviser to Fukushima Prefecture. Hence, he had been involved in nuclear accident management twice before becoming NRA Chairman. As head of the NRA he proved determined to prevent another accident by pushing for better safety governance.

Another noteworthy choice for the board was Ōshima Kenzō due to his recurring encounters with the terrible side effects of radiation and risks related to nuclear power. Born in Hiroshima in 1943, he was a survivor of the atomic bombing and the "black rain" that followed. Later, during his time as United Nations Undersecretary General for Humanitarian Affairs and United Nations Coordinator of International Cooperation in Chernobyl in the early 2000s, he visited the contaminated areas around the Chernobyl nuclear power plant and was instrumental in the compilation of a report titled "The Human Consequences of the Chernobyl Nuclear Accident: A Strategy for Recovery" (United Nations 2002). Furthermore, in 2011–12, Ōshima was a member of the NAIIC and therefore part of the investigation committee that came to the conclusion that Japan's pre-3.11 nuclear safety governance system was inherently flawed and in need of a major overhaul.

Even before the NRA's inauguration and the board's official appointment, designated board members and NRA officials began working on ensuring the longevity of their vision of independent safety governance. As cabinet advisers, designated board members joined the team in charge of making arrangements for the NRA. This preparation office was headed by the MOE official Morimoto Hideaki, who would soon after become the Deputy Secretary General, the second-highest position in the NRA Secretariat. The team also included the NRA's first Secretary General and a few other members who would go on to work in senior positions within the NRA. Their task was to translate a piece of legislation, the NRA Act, into a functioning nuclear safety agency. They drafted organizational guidelines to be adopted at the inaugural meeting (interview with NRA official, 2018, August 14). At the NRA inaugural meeting on September 19, broadcast live on YouTube,[7] the board adopted transparency guidelines for NRA board meetings, guidelines for managing administrative documents, and a neutrality code of conduct for NRA board members (NRA 2013b, 14), which remain at the heart of NRA operations.

The first board, and Chairman Tanaka in particular, stressed transparency from the start. At a preparatory meeting, held at the MOE a week before the inauguration, draft guidelines were tabled and everybody present agreed to introduce one of them, transparency, into the NRA Law (MOE 2012). Implementing transparency even before the inauguration, they invited the press to observe their meeting and later published their first video recording. The decision to be as transparent as possible was made early on when designated board members and senior officials sought to achieve regulatory independence in line with international norms. Transparency was considered to be an important aspect to make the NRA independent not only legally but also in its day-to-day operations (interview with Former Board Member, 2019).

Once appointed, board members worked with the Noda Cabinet to introduce some guidelines into the legal framework as supplementary provisions. These included transparency, staggered terms of office, and the abolition of JNES so that the NRA could take over its functions. Since JNES was in charge of monitoring safety at nuclear power plants, integrating it into the NRA was equivalent to the new safety regulator acquiring monitoring competencies. These additions to the NRA Act ensured the longevity of these reformers' understanding of independent and transparent safety regulation and became cornerstones for the NRA's work.

Transparency guidelines and guidelines for administrative documents interlocked to form an information release system not subject to request. All decisions and procedures should be written down and made publicly available and all NRA board meetings should be open to the public. In January 2013, a high-ranking NRA official breached the guidelines by handing over an unreleased report to a nuclear operator prior to a meeting. In response, the NRA made the incident public, issued a written warning, and transferred the director general for nuclear regulation back to the Ministry of Education, Sports, and Science (MEXT), where he had come from (NRA 2013b, 58). To avoid future incidents, the NRA Commission amended transparency guidelines soon after to prevent NRA officials from meeting alone with nuclear operators. Such meetings and the topics discussed were henceforth subject to public disclosure (NRA Commission 2013b). Immediately transferring a first-time offender set a warning example and showed the NRA board's determination to enforce transparency.

With a code of conduct for commission members, the NRA Commission braced itself for the appointment of commissioners with close ties to the nuclear industry. Referring back to Article 11 of the NRA Act, which required information disclosure for board members, the board's code of conduct for commission members contained three main rules: 1) board members are prohibited from receiving donations from nuclear operators during their term in office, 2) they must disclose any income received in the three years immediately prior to assuming office, and 3) they are asked to disclose incidents where a student of theirs finds a job at a nuclear operator (NRA Commission 2012a). These rules resemble the selection criteria spelled out by the DPJ. The advantage of this binding code of conduct became apparent when the LDP ignored the DPJ criteria during the appointment of a board member in 2015, but NRA rules forced a disclosure of the information the LDP had held back in the selection process. Hence, should the government appoint an expert with close ties to the industry, it would at least be publicly known, and the new board member would have to sever those ties.

NRA guidelines and the mission statement reflected the lessons of the 3.11 nuclear accident that the NAIIC spelled out. To achieve the overarching goal of regaining public trust in nuclear safety, the mission statement, adopted in early 2013, listed four concrete problems, identified by the NAIIC, that the NRA as an organization sought to avoid: 1) regulatory capture as a result of lacking expertise within the regulator, 2) delayed implementation of relevant regulation, 3) making regulation partly voluntary for electric utilities, and 4) a lack of independence from the political arena, METI, and the electric utilities (NRA 2013b, 1). To achieve this, the NRA board pledged to make the NRA transparent, open to advice, and grounded in independent and effective decision-making, including its own data collection (NRA Commission 2013a).

A central task in 2013 was the finalization of safety standards. To begin with, the NRA adopted emergency response guidelines in February 2013. A first draft was revised following public comments. Concretely, provisions for stocking iodine were expanded following numerous public comments to that end (NRA 2013b, 51–53). Shortly after the Abe government had officially approved the preliminary first board in February 2012, the board adopted the exact safety regulation the LDP had prevented from being introduced into the NRA Act. The NRA board adopted a forty-year rule for the operation of nuclear power reactors, which allowed for a one-time extension of twenty years provided the reactors pass an additional safety screening. It combined the forty-year rule with a backfit system to ensure that nuclear reactors would be updated in line with changes in safety requirements (NRA 2013b, 29). Draft safety regulations were presented to the public for comments beginning in April, and final safety standards were issued in June and went into force on July 8, 2013. Adopting new safety requirements established the basis for nuclear reactor safety checks (discussed in more detail in chapter 4). In particular, the backfit system marked a major change in safety governance, because for the first time since the beginning of Japan's nuclear power program, regulators acquired the competency to force utilities to make safety improvements after the initial operating license had been granted.

Other early actions pertained to the damaged Fukushima Daiichi nuclear power plant (NRA 2013b, 14). Monitoring the damage at this plant was not a task outlined in the NRA Act, but this was quickly taken on by the regulatory agency. The six core tasks of the NRA, as listed in its annual "NRA Policy" document, are 1) restoring trust based on independence, neutrality, and transparency; 2) strict implementation of safety regulation; 3) overseeing measures taken to decommission the damaged Fukushima Daiichi nuclear power plant; 4) constructing a scientific knowledge and human resource base for ensuring nuclear safety; 5) implementing nuclear security and safeguards; and 6) establishing solid emergency response measures and monitoring policy (NRA Commission 2015b, 2019). While the latter three tasks were prescribed by the NRA Act, the first three were assumed by the NRA board soon after its establishment, which demonstrates the independence with which it molded nuclear safety governance according to its own vision.

Political Influence via Board Members' Appointments?

For politicians, the easiest way to gain influence over regulation would be to appoint board members with close industry ties and receptiveness to political and industry pressure. Technically, there was a chance to replace the whole first board in early 2013, because it still required Diet approval. Instead of replacing it, the Abe government ex post facto approved the board in February. Replacing the NRA board immediately after coming into government would have undermined the credibility of the LDP's commitment to improving nuclear safety in the face of a critical domestic and international audience by establishing an independent regulatory agency, a commitment the LDP repeated like a mantra throughout the reform process in 2012, discussed in chapter 2.

Staggered terms of office for board members ensured continuity but also offered the opportunity to install someone with government-aligned views each time there was a new appointment, in 2014, 2015, and 2017. The most important appointment decision came in 2017, when Chairman Tanaka, age seventy-two, resigned at the end of his term. Tanaka was followed by Fuketa Toyoshi, who had been on the NRA board since the beginning, clearly understood the aims Tanaka Shun'ichi had pursued as chairman and was instrumental to enforcing thorough safety checks as commissioner for nuclear safety. While both Tanaka and Fuketa had a history of working for the JAEA, which has been widely criticized for its involvement in collusion within the nuclear village, they showed a strong willingness to foster safety governance based on independence, transparency, and neutrality. With Fuketa's appointment as chairman, the safety agency passed the litmus test of withstanding political influence via the appointment process. The selection process is not public, but chapter 5 attempts to shine light on the interactions between the NRA, the MOE, and Prime Minister Abe in the process. Instead of a political appointee to bring the agency more in line with government views, someone from within the board was elected to continue the path the NRA had set out on.

Another potential avenue for political influence is the budget process. Its legal status as an Article 3 organization grants the NRA the right to draft its own annual budget request. These budget requests are deliberated as part of the government budget-making process, like that of every other agency or ministry. The NRA's actual budget increased concurrently with the expansion of tasks and staff numbers in the first two years. In FY 2012, it was ¥37.76 billion (NRA 2013b, 6), which rose to ¥63.2 billion in FY 2013 (NRA 2014a, 2) and FY 2014 (NRA 2016b, 30). For future years, the NRA expected little change in the overall amount but rather changes in the internal allocation of funds to meet its needs in the face of changing tasks (interview with NRA official, 2016, March 14). This expectation was confirmed by the FY 2017 and FY 2018 budgets, which were ¥57.4 billion (NRA 2018, 78) and ¥59.1 billion (NRA 2019a, 104), respectively. The budgets allocated seemed sufficient. This allowed the safety agency to undertake its activities and, for a few years, even provided for more positions than the NRA could fill. Unlike its predecessor NISA, which was financially dependent on METI, the NRA was as financially independent as possible within the Japanese administration with no signs of political tampering.

In sum, the first NRA board had a clear vision of Japan's post-3.11 nuclear safety governance, which aimed to remedy the flaws identified by the NAIIC, included top-down safety regulation opposed by the LDP during the reform process, and put a strong emphasis on transparency and accountability. Working with the outgoing DPJ government at first and then using its status as an independent Article 3 organization, the first board around Chairman Tanaka and senior NRA Secretariat officials swiftly enshrined their vision into binding rules and supplementary provisions in the NRA Act in order to ensure its longevity.

The Secretariat

The NRA board is supported by a "secretariat" (*kiseichō*). Over the course of roughly three years, it was formed as a jigsaw puzzle, with each piece consisting of functions and related staff from elsewhere. Note that the terms agency and secretariat are used interchangeably in this chapter. Initially, the NRA took over the tasks of NISA and the NSC, namely ensuring nuclear safety and security. In April 2013, the tasks of radiation monitoring and safeguarding radioactive material, as well as related staff, were transferred from MEXT to the NRA, increasing the number of full-time staff from 473 to 527. In March 2014, the NRA took over JNES's staff and functions, such as undertaking on-site inspections of nuclear reactors, providing advanced vocational training for nuclear experts, and conducting research on nuclear safety standards. As a result, the number of full-time staff increased to 911 (NRA 2013b, 7; interview with NRA Official, 2016). Thus, within less than three years, the NRA Secretariat doubled in size to become an agency with about one thousand full-time employees, roughly half the size of the Ministry of the Environment. Figure 3.2 displays this process by showing where functions and staff integrated into the NRA came from (ovals) and how it affected the size of the NRA (rectangle).

To accommodate the different tasks, the agency was reorganized repeatedly. In 2014, it consisted of three technical departments and the Secretary General's Secretariat, an administrative department. The administrative department was in charge of public relations, human resources, budget and accounting, and international affairs. The three technical departments were a regulatory standard and research department, a radiation protection department, and a nuclear regulation department. These departments were supplemented with a Human Resources Development Center, twenty-two regional offices, and local radiation monitoring offices (NRA 2014a, 7). Later, the Secretary General's Secretariat was expanded to include a division of legal affairs—to deal with surging numbers of

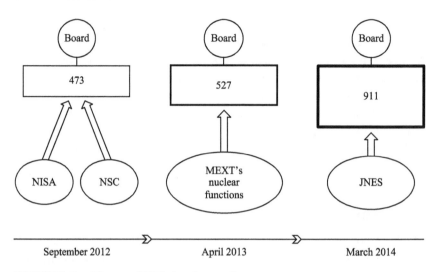

FIGURE 3.2. Stages of NRA development

lawsuits against NRA decisions—and the Nuclear Regulation Department was divided into licensing and oversight divisions to accommodate the introduction of a new on-site inspection system (NRA 2019a, 11). With that, the NRA Secretariat took on its final shape for the time being.

Medley of Competencies

The NRA Act (Article 4) mandated that the NRA ensure the safety of nuclear facilities and the security of nuclear material, safeguard nuclear material based on international commitments, supervise radiation monitoring, investigate the causes of the Fukushima nuclear accident, train researchers and engineers, formulate emergency preparedness guidelines, and prevent radiation hazards by regulating radioisotopes used outside of nuclear reactors. Also, the NRA shares jurisdiction over the National Institute of Radiological Sciences (NIRS) and the Japan Atomic Energy Agency (JAEA) with MEXT.

After taking over these tasks one by one, the NRA was in charge of a medley of competencies. Regulatory competencies break down into three steps: rule-making, monitoring, and sanctioning competencies. The only area fully under the NRA's control is that of nuclear safety. The NRA Act put rule making for nuclear facilities into the hands of the NRA. Integrating JNES and its monitoring functions into the NRA also put it in charge of monitoring implementation. The backfit system introduced by the NRA board could be used as a sanctioning mechanism by the NRA threatening to revoke licenses unless all safety standards are met. Neither monitoring nor sanctioning was part of the legal framework. The NRA used the power granted by the NRA Act to adopt top-down regulation and sanctioning power without negotiating with the industry or consulting the government mere weeks after the Abe government officially approved the first board. By proactively assuming monitoring and sanctioning power, the NRA came to be in charge of all three steps of nuclear safety regulation.

In the area of nuclear security and safeguards, the safety agency translated provisions from international treaties Japan has signed into domestic regulation and oversaw the implementation. Regarding radioactive materials and radioisotopes, a division of labor between the NRA and the Ministry of Health, Labor, and Welfare (MHLW) emerged. The MHLW remained in charge of devising rules for radioactive materials and radioisotopes with regard to human health. The NRA regulated the handling of such materials, for example in research or medical treatments. Similarly, the NRA devised rules for radiation monitoring and oversaw implementation by the different agencies and organizations involved in radiation monitoring.

For the cleanup operations at the Fukushima power plant, the NRA devised rules and began monitoring their implementation. Without sanctioning powers, such as decommissioning permits, the NRA lacked a lever to enforce its rules other than to publicly name and shame TEPCO in case of violations. Hence, the NRA monitored the process of decommissioning the damaged Fukushima Daiichi nuclear power plant rather than actively shaped the process.

Regarding emergency response measures, rule-making competencies resided with the NRA, but implementation and sanctioning competencies were taken over by the Cabinet Office. This was possible due to a vague phrase in the NRA Act. In 2012, a point of contention between the DPJ government and the LDP/Kōmeitō opposition had been the prime minister's role in emergency response management. The compromise they found put the NRA in charge of technical aspects of emergency response while a Nuclear Emergency Preparedness Council (NEPC) at the Cabinet Office was to oversee operational aspects. At first, the NRA and the Cabinet Office shared emergency response tasks. With the integration of JNES into the NRA, the NEPC was established in the Cabinet Office as well as a supporting emergency preparedness and responses division. Out of the sixty to seventy officials working inside the emergency preparedness and responses division in support of the NEPC, about forty were actually NRA officials dispatched to the division (interview with NRA Official, 2015). With the NEPC's establishment in 2014, the Cabinet Office took over the monitoring competency regarding emergency response plans and the ability to sanction violations by withholding consent to a proposed evacuation plan.

In short, the NRA Secretariat grew quickly in the first two years, even outgrowing initial plans for an agency of about five to six hundred people (Nihon Keizai Shimbun 2011), by taking over a medley of competencies related to nuclear safety, security, and safeguarding. It has the strongest regulatory competencies in the area of nuclear safety, where the NRA proactively assumed monitoring and sanctioning powers. All decisions, be they regulatory or otherwise in nature, are made using what is dubbed the "council system" here.

The Council System

NRA decision-making follows a process best described as a council system, because all reporting and deliberating take place in councils. Apart from board meetings, four main types can be distinguished: study teams, expert meetings, councils, and committees. Figure 3.3 provides an overview of the council system. Note that dashed lines signify temporary councils established by the NRA Commission, as opposed to permanent councils written into the NRA Act, which are represented with continuous lines. Ovals overlapping with the secretariat represent councils that include secretariat officials as members. If board members participate in a council, a double arrow connects the board and the respective council.

The council system basically created a two-step process. In the first step, reports and information are provided in a specific council. After deliberations within the council, representatives from that council present their case to the NRA board. Also, department heads, members of regional offices, and staff dispatched to the Cabinet Office regularly provide accounts of recent activities and ask for approval of plans during board meetings. Based on reports received and, if applicable, additional information or amendments requested, the board makes final decisions on all matters. Hence, it is a system in which everyone reports

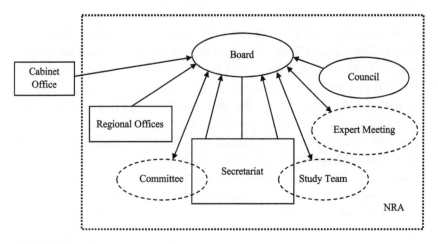

FIGURE 3.3. The council system

back to the NRA Commission. The council system allows the safety agency to deal with a number of issues in parallel despite all final decisions resting with a board of five scientists.

Officially, the board's decision-making procedure is based on voting among the five members. In case of a tie, the right to make a decision lies with the NRA chairman (Article 10 of the NRA Act). In practice, no actual voting is observable at commission meetings. Instead, Chairman Tanaka established a consensual decision-making process. Each commissioner gets the chance to ask questions for clarifications and to require additional information about the topics raised. If a topic seems sufficiently discussed, the chairman asks about any objections. In case of no objections, a decision is adopted. The board making decisions unanimously means no board member's opinion on a topic can be disregarded. Giving every board member a veto can limit the influence of political pronuclear appointments in the board. Even if only one or two critical board members were left, they would have the chance to push for a stringent interpretation of safety standards.

Between October 2012 and October 2013, four study teams[8] prepared a safety review system and new safety standards for commercial power plants as well as nuclear fuel cycle facilities. Another three study groups deliberated emergency response measures, such as stocking medicine, raising preparedness levels, and monitoring emergency preparedness between November 2012 and August 2015. As of 2021, five of the ongoing study teams were in charge of further developing safety standards and regulatory oversight, four dealt with decommissioning and radioactive waste, two focused on emergency preparedness measures and medicine, two dealt with monitoring and evaluating volcanic activities, and one looked into cybersecurity.

Two permanent study teams were in charge of reviewing the safety of commercial power plants and fuel cycle facilities, called review meetings. They were held for the eight hundredth time in November 2019 and for the three hundredth time in September 2019, respectively. Prior to review meetings, NRA Secretariat members and commission members review the documents submitted by electric utilities. Representatives from the utilities operating the reactor in question are invited to present their case and to answer questions. Decisions made during review meetings are later presented to the NRA Commission, which can either approve the safety of a reactor or demand more information and changes to refurbishments, in which case it goes back to the study team.

Committees are similar to study groups but consult external experts more extensively. Which experts may be consulted is spelled out in the guidelines on the neutrality of external experts. Excluded are those who have held a position in the relevant electric utilities in the previous three years, have personally received

¥500,000 or more in remuneration from the relevant electric utility during one fiscal year,[9] or have been involved in prior examinations of the facilities in question (NRA Commission 2012c). Similar to the guidelines for selecting board members, NRA guidelines excluded experts with close ties to the industry and those who were part of previous investigations. Avoiding experts close to the industry proved difficult, as the example of Tanaka Satoru showed. Prior to his appointment as a board member, he had been consulted as an external expert. While NRA guidelines require such experts to report funding they had received from the nuclear industry, he failed to disclose all funding. Even without foolproof neutrality guidelines, insistence on hearing previously uninvolved experts provided an opportunity for voices that were not heard previously and brought new knowledge into the safety discussion.

Committees cover a variety of topics, including assessments of damaged nuclear power plants, analyzing the causes of the Fukushima Daiichi nuclear accident, addressing nuclear security issues, and dealing with aging reactors and nuclear waste issues. Early on, the NRA conducted earthquake fault line investigations for six nuclear facilities, which the NRA's predecessor NISA was unable to conclude due to its dissolution. For each investigation it held expert meetings.[10] In case of the Tsuruga NPP, the expert meeting not only conducted on-site visits but also engaged in excavations in order to gather its own data on past fault line activity. Expert meetings sought technical input on a number of issues, such as plant safety, nuclear fuel and material, severe accidents, nuclear fuel cycles, and radioactive waste, as well as earthquakes and tsunamis. Another set of expert meetings was dedicated to reviewing organizational practices and policies of the NRA.

The NRA Act mandated the establishment of standing councils to ensure an external evaluation of NRA decisions and regular input by external experts. Secretariat officials attend to observe and to report about efforts made to meet expert recommendations, but board members rarely attend. Rather, standing councils extensively consult external experts in line with the neutrality criteria. There is one council on reactor safety and nuclear fuel safety examination, one on the prevention of a radiation hazard, and one on national research and development agencies, such as NIRS and JAEA. Standing councils exist to ensure a continuous critical evaluation of the NRA's practices and regulations by outside Japanese experts.

Tracing the issues covered in councils reveals the NRA's priorities at any given time. In the beginning, the NRA focused on establishing safety standards, a system for safety checks, and emergency response measures. Since then, the range of topics covered has expanded, most notably to include aging reactors, decommissioning, and nuclear waste storage issues. At the same time, a number of councils were dedicated to improving safety standards, including cybersecurity and terrorism as new areas, revamping the safety inspection system, and evaluating the NRA's progress as an independent safety agency. Hence, next to recurring themes—safety regulation, inspections, and agency (self-)evaluation—increasing attention was paid to dealing with an aging set of reactors and all the issues that come with it.

Unprecedented Transparency, But Limited Public Participation

A new information release system, which was not subject to disclosure requests, was formed by interlinking guidelines for transparency and administrative documents. The NRA adhered to administration based on written documents, in order to document the discussions leading to a decision and to disclose discussion and related documents (NRA Commission 2012b). An exception to this rule was meetings about safeguarding nuclear material for reasons of national security.[11] NRA transparency included streaming council meetings and press conferences live on the NRA's YouTube channel, "NRAJapan." Reference materials used during meetings were published on the NRA's website beforehand and meeting minutes afterwards. Together, the NRA's website and YouTube channel became a repository of written documents and video captions.

For board meetings, there was even an option to attend meetings in person at the NRA building in Roppongi, Tokyo. This option was not confined to Japanese citizens but could be made use of by anyone with a valid passport (and sufficient Japanese language knowledge). Together with video recordings and openly available written documents, these practices allowed the public to literally watch the NRA at work, to follow discussions, and to understand how a topic was raised successively in different meetings and why a board decision was made.

Regarding the media presence, the NRA renounced the system of preselected accredited "journalist clubs" usually used by ministries and agencies. This system has been criticized for providing major media organizations preferential access to government information and for creating a pro-elite bias in reporting (Freeman 2000). Rather than relying on "journalists clubs," board meetings and press conferences were opened to all media, including freelancers and nonaccredited journalists.

One concern was whether transparency could also put an end to backroom meetings between industry representatives and members of the nuclear safety administration. Early on, this became an issue when an NRA official handed over an unpublished report to a utility before a meeting in January 2013. In response, the board made an example of the official and amended transparency guidelines to cover meetings with regulated parties. It decided that they should be held by at least two NRA officials and that the interview schedule and reference materials would be disclosed (NRA Commission 2013b). Disclosing the when, who, and what of such meetings was, of course, no guarantee that backroom agreements would completely be a thing of the past, but this certainly made them more difficult.

An issue that remained was the difficulty of understanding discussions. Reference materials containing technical files with design specifics for nuclear reactors, seismological data, NRA staff data, and so on, are practically incomprehensible without prior technical knowledge or the time for extensive research. This led Ban Hideyuki, head of Japan's biggest antinuclear NGO, the Citizen Nuclear Information Center (CNIC), to criticize NRA transparency practices for creating an information overload that requires much time and energy to follow and to unpack (CNIC, pers. comm.). A senior NRA official showed awareness of the "information pollution" criticism and the lack of explanations. But he explained that the safety agency was lacking the staff and time to provide that: "Transparency can mean many things, such as openness, understandability, and accessibility. What we can achieve right now, is transparency as openness, not understandability, not accessibility" (interview with NRA Official, 2016). Being open and transparent for the NRA was an important action in order to regain public trust and to let the public evaluate its work, but this has not translated into easily accessible information.

In the late 2010s, the NRA began to reach out more to explain its work. In a guideline adopted in 2017, the NRA board pledged to visit more nuclear power plants under review to engage in conversation with local stakeholders (NRA 2019a, 5). The same year, there were eighteen meetings where NRA officials exchanged opinions with local governments. In addition, NRA officials, and former Chairman Tanaka, attended over one hundred meetings with local stakeholders,[12] from residents to local politicians, experts, and heads of local governments, in order to answer questions and explain the NRA's work (NRA 2019a, 131–36). Rather than through its website or YouTube channel, the NRA chose to engage directly with local stakeholders to increase transparency and accessibility in municipalities affected by nuclear restarts.

To increase transparency toward non-Japanese speakers and to reach a broader international audience, some documents were translated into English, including annual reports and announcements on its website. The board also published the documents submitted in English for the IAEA IRRS review. This once again reflects that international actors remained a key audience after reforms.

Transparent practices allowed for considerable public scrutiny. However, communication with the public remained a one-way street. While the NRA provided channels for public feedback, it did not create a two-way street of public participation in the process. While there was no direct citizen, or NGO, participation, the NRA exhibited a very high degree of transparency and began reaching out to local stakeholders.

Such a high degree of transparency also had an effect on NRA officials. For them, transparency meant working under constant public scrutiny with every word spoken in council meetings immediately becoming public knowledge. It was not easy for the NRA board and officials to make all decisions under public scrutiny and pressure from the public (interview with NRA Official, 2015). As an official described it: "Everybody was nervous at first, working in front of cameras and an audience at the back of the room, that would sometimes shout out their disapproval of ongoing discussions" (interview with NRA Official, 2018). Intended to open the NRA to the public, transparency also kept the NRA board and officials on their toes by reminding them that the public they had pledged to protect was watching.

Transparency has played an important role for the NRA's ability to persist in the face of pressure. In fact, it produced results beyond expectations, even those of former chairman Tanaka. He felt it increased the NRA's reliability in public perception because it was clear to everyone how a decision came about. For the same reason, transparency also helped decrease opposition as it was more difficult to challenge a decision that had been explained and made in the open. In that sense, it also helped protect the NRA's independence. Speaking about this, a former board member expressed a sense of pride about the NRA being highly transparent, even in international comparison, and called for more government bodies to adopt the same approach (interview with Former Board Member, 2019). Similarly, the second chairman, Fuketa Toyoshi, pointed to transparency as one factor protecting the NRA's independence (NRAJapan 2017).

External Evaluation by the IAEA

Not long after its establishment, the NRA board sought external evaluations by the IAEA. In 2014, the NRA board requested an IAEA Integrated Regulatory Review Service (IRRS) Mission. Such IRRS missions assess a regulator's activities against the IAEA standards for independent safety regulation. Based on the NRA's Self-Assessment of Regulatory Infrastructure for Safety (SAIRS) report of almost two thousand pages and visits to the NRA, the IRRS Mission presented its report in April 2016. The IAEA assessment lauded two good practices and gave a total of twenty-six recommendations and suggestions. Good practices mentioned were a) the establishment of an independent and transparent regulatory body with increased powers and b) incorporating the lessons learned from 3.11 into the legal framework. In its recommendations and suggestions, the IAEA highlighted three issues in particular: first, to improve the nuclear safety inspection system;

second, to further enhance human resources through hiring experienced staff and through training, research, and international exchanges; and third, to promote a safety culture and a questioning attitude (IAEA 2016c, 2).

In response, the NRA set out to overhaul the three areas of its work for which the IAEA recommended improvements. Regarding the safety inspection system, the NRA established a Study Team on Revising the Inspection System and sought further input from the public and the Standing Council on Reactor Safety and Reactor Fuel Safety Examination. In early 2017, the NRA board approved a draft bill to revise the Act on the Regulations of Nuclear Source Material, Nuclear Fuel Material and Reactors and subsequently submitted it to the Diet for approval (NRA 2018, 15). This became the legal foundation for a revised inspection system that adopted a wider scope of inspections and granted more powers to inspectors.

To enhance human resources, the NRA's Human Resources Development Center revised the two-year training program for technical experts. It refined necessary qualifications for safety reviewers and inspectors and introduced a new licensing system for nuclear safety inspectors based on an examination at the end of the training course (NRA 2018, 60). Regarding a safety culture and a questioning attitude, Chairman Fuketa stressed in his first press conference that this presented one challenge the NRA needed to tackle. In practical terms, this meant integrating a questioning attitude into revamped safety inspections. In addition, the NRA stressed the need for a shift in safety culture towards plant operators proactively seeking to improve nuclear reactor safety. As Chairman Fuketa put it, the NRA sought to ingrain a "sense of my plant, my responsibility" (*mai-puranto-ishiki*) (NRAJapan 2017). The safety inspection revision and safety culture issues are discussed in chapter 4 in more detail.

Not long after, in August 2017, the NRA requested an IRRS follow-up mission to assess its progress, which took place in early 2020, once the new inspection system was fully up and running. The IAEA follow-up report concluded that the NRA had successfully addressed twenty-two of the twenty-six recommendations and suggestions made in 2016, including the improved inspection system and enhanced staff qualification and training programs. Establishing a safety culture at nuclear power plants around Japan was, however, mentioned as an area that required continued efforts. Furthermore, the IAEA report pointed out the NRA's increased participation in the global nuclear safety regime and encouraged the government to provide the NRA with sufficient resources to continue its international engagement. The report also lauded the extensive openness and transparency of NRA in the review process (IAEA 2020b). Overall, the IAEA follow-up mission concluded that the NRA had quickly amended its practices to ensure compliance with international norms for independent nuclear safety governance and had become ever more engaged in exchanges with international peers and the global nuclear safety regime.

International Engagement

Through various avenues, the NRA pursued cooperation and exchanges with international counterparts. It entered numerous bilateral information exchange cooperation agreements, including with nuclear regulation agencies in the United States, France, the United Kingdom, Sweden, Russia, Spain, Turkey, Vietnam, Lithuania, and Germany. On a trilateral basis, South Korea, China, and Japan agreed to do the same at a China-Japan-Korea Top Regulators' Meeting (TRM) in November 2013. The NRA entered a monitoring cooperation with the IAEA, where IAEA members joined NRA officials in seawater sample collection to monitor marine radioactivity and in reviewing the progress made in decommissioning the Fukushima Daiichi NPP. NRA representatives also actively participated in a number of international conferences (NRA 2014a, 17–18, 94–95).

International exchanges also played a role in training staff, who were sent abroad for practical training, for example with the IAEA, the US NRC, or the Nuclear Energy Agency at the OECD (NRA 2014a, 14; 2015a, 2016a). Such exchanges, of course, require proficiency in English, which officials could acquire as part of the in-house training for NRA officials, discussed below. The English proficiency levels of the NRA's first chairman, Tanaka Shun'ichi, were rather low, but his successor Fuketa Toyoshi appeared to be fluent enough to easily exchange views with international counterparts. In fact, Fuketa was considered a good successor not only due to his technical expertise but also because of his English skills (interview with Former Board Member, 2019). Overall, this shows a substantial willingness on the part of the NRA to engage with international counterparts.

In-House Expertise

It is essential for a regulating agency to have in-house expertise and competent staff in order to be independent. There were fears that practices of industry-friendly regulation would live on, either through the NRA taking over staff from its predecessor organizations NISA, the NSC, and JNES, or through METI regaining influence via interministerial staff dispatches. Aware of this danger, the first board made it its mission to avoid regulatory capture as a result of lacking expertise (NRA 2013b). NRA efforts to ensure sufficient competent staff and in-house expertise included a "no-return rule," a "no-*amakudari* rule," staff diversification through hiring, and vocational training.

Initially, NRA officials consisted primarily of people transferred from NISA, the NSC, MEXT, and JNES. As of late 2015, the NRA had a capacity of 968 full-time positions, of which 51 were vacant (interview with NRA official, 2016, March 14). Out of about 900 full-time staff members, 250 had come from METI, 115 from MEXT, 20 from the MOE, and 35 from other agencies such as the Police Agency or the Fire Department. Hence, 420 people originally transferred from other parts of the bureaucracy remained within the NRA. Furthermore, there were about 300 people originally from JNES working for the NRA. Thus, after three years, about 45 percent were from other parts of the administration. Another 30 percent were former JNES employees, and the remaining 25 percent were hired by the NRA.[13] By early 2018, the share of staff hired by the NRA had increased to 34 percent.

The no-return rule applied to officials within the NRA who were initially transferred from other parts of the administration, either as part of the reform or as temporary dispatches. They could not return to divisions, departments, or positions directly involved in the promotion of the use of nuclear energy, which can be found in ANRE, MEXT, and the AEC in the Cabinet Office (NRA 2015b).[14] Not including whole agencies or ministries in this rule left a loophole via subsequent transfers within ministries. The NRA attempted to close this loophole with the first draft of the no-return rule, which included a passage about preventing those leaving the NRA from taking up any of the banned positions even later on in their career. But such decisions lie with the human resource division of the respective administrative body, leaving the NRA no means of ensuring compliance (NRA Commission 2015a, 33). Interestingly, this first draft was presented by a senior official in human resources who had previously worked for METI, precisely the ministry that used to have the biggest influence over safety regulation staff. As a result of the no-return rule, officials from other ministries were prevented from immediately returning to departments or positions promoting the use of nuclear power, but they could still do so in the longer term. This rule created a hurdle for returning to such positions, which rendered a temporary transfer to the NRA a less attractive option.

The suspension of the no-return rule for the first five years elicited criticism as a door opener for influence by pronuclear parts of the administration. The decision was made to accommodate career patterns of officials who were transferred into the NRA as part of the reform. During the first five years, they were treated as dispatches from the ministries they originally came from and were given a choice to join the NRA or to return to their original ministry. Between September 2012 and September 2015, almost three hundred officials left the NRA and returned to their home ministries, including one person returning to a post within METI and seven to posts within MEXT covered by the no-return rule (NRA Commission

2015a, 35). Regarding the likelihood of dispatches remaining within the NRA, an official in human resources expected some of the officials from METI and MEXT to be eager to return to their home ministries, as those were the ministries they initially chose for their career. Other people seemed willing to stay, and people in positions of a director or higher were mostly expected to remain within the NRA (interview with NRA Official, 2015). Despite its suspension for five years, it appears the no-return rule mostly inhibited a direct revolving door between posts in regulating and posts in promoting nuclear power.

Furthermore, NRA officials who are temporarily dispatched to other ministries were banned from joining nuclear policy divisions within METI or MEXT (interview with NRA Official, 2015). Officials dispatched to the NRA from other parts of the administration were mainly used for managerial tasks, such as budgeting. Positions in areas that require a lot of knowledge and expertise, such as safety inspector, examiner of safety review documents, and emergency response official, were staffed with NRA officials only (interview with NRA Official, 2018). Hence, the aim to limit external influence was also apparent in the NRA's own staff allocation and staff dispatch practices.

Next to interministerial staff exchanges, the NRA also addressed the revolving door with the nuclear industry. The no-*amakudari* rule means NRA staff may not take on retirement positions within any of the regulated companies (interview with NRA Official, 2015). The *amakudari* phenomenon is prevalent within the Japanese administration and is closely related to the way people are promoted within organizations of the bureaucracy. An "up or out" mechanism results in many officials leaving in their mid-fifties and taking up industry positions to bridge the gap until they reach retirement age. The NRA took in a number of senior people in the first years, especially from JNES. In 2015, 172 officials (out of about 900) were fifty-eight years old or older (interview with NRA Official, 2016). Due to the need for skilled staff, the NRA employed many of them past the standard retirement age of sixty, either through temporary full-time contracts or as part-time staff. In addition, the NRA adopted later retirement ages of sixty-three and sixty-five for highly specialized officials, such as safety inspectors and examiners of safety review documents. And if they chose to, they had the opportunity to stay as part-time staff even beyond that (interview with NRA Official, 2018). With these rules and many seasoned NRA staff past the *amakudari* age and still urgently needed, *amakudari* is unlikely to take hold in the near future.

Hiring was an important task from the beginning to fill vacant positions as officials returned to their home ministry or retired. Attracting experienced staff, however, was not easy due to competition from the industry. While the NRA was looking for experienced staff to become regulators or inspectors, electricity utilities were looking for experienced engineers to refurbish nuclear reactors in order to pass safety checks (interview with NRA Official, 2015). By April 2018, the NRA had hired a total of 219 experienced staff and another 118 fresh graduates. In total, the NRA employed 984 people full time by then, rendering it almost fully staffed (NRA 2019a, 70). By that time, staff hired by the NRA made up 34 percent of the total. The need to hire presented a challenge but also an opportunity to shape an entire new generation of nuclear safety regulators.

Crucially, the NRA created a career path for nuclear safety regulators that did not exist before. Prior to 3.11, many of those working in nuclear safety regulation were dispatched from other parts of the administration or the industry and would leave regulation agencies after a few years to pursue their careers. Thus, the NRA establishing a career path as a nuclear regulator was a major change. Officials working for the NRA must choose to embark on a career as a regulator. For the necessary training, a Human Resources Development Center (HRDC) was established in 2014. The HRDC developed different model career paths: generalists, experts, and research specialists (NRA HRDC 2015, 10). Generalists acquire only basic knowledge of nuclear power as they later go on to work in coordinating and policy planning positions. Experts require a specialized skill set in one area, such as safety review, on-site inspections, or emergency preparedness. Research specialists choose a field of research to specialize in and acquire as much knowledge of it as possible.

In response to IAEA recommendations to strengthen human resource development, the NRA revised the burgeoning training system in 2016. In the process, it clarified the competencies and model career paths, for example for safety inspectors. The IRRS had pointed out a need for better safety inspector training. To learn from abroad, the NRA sent a study team to the US NRC to study its inspector training system in July 2016 (NRA 2018, 60). Soon after, it introduced a revised, more fine-grained, two-year training scheme that included separate training plans for safety inspectors, nuclear safety reviewers, safeguards inspectors, emergency preparedness officials, and radiation regulators (NRA 2019a). A major change was the introduction of an accreditation system for inspectors, for which they have to pass examinations at the end of their training. Under the new training program, the NRA enlisted the help of at least one inspector trainer from the NRC to hold seminars and share experiences from the United States with Japanese inspectors (interview with NRA Official, 2018). Another change was the installation of two training simulators inside the NRA building, one resembling older manual reactor controls and one using digital controls for newer reactors, to prepare staff for the task of regulating safety as well as to comprehend and respond to emergencies (NRA 2018, 59). Other things remained unchanged, such as the first six months of joint basic training for all career paths, which seeks to instill the NRA's missions and practices in all its employees.

In April 2018, the new training program was up and running. That year, five officials started taking the two-year full-time training program. In April 2019, about twenty fresh graduates were scheduled to start their training (interview with NRA Official, 2018). Since most NRA officials were previously employed elsewhere, the NRA offered graded training for fresh graduates, midcareer recruits, section chiefs, deputy directors, and administrative positions, depending on training needs. For midcareer hires, there were courses on specific tasks, such as training for nuclear safety inspectors, physical protection inspectors, and radiation inspectors. Graded refresh training was aimed at filling the knowledge gaps of the staff taken over from other parts of the administration or hired midcareer. But the new training program offered the chance to provide structured training (interview with NRA Official, 2018). For the purpose of international exchange, the HRDC also offered training to enhance the staff's communication potential through providing different levels of English classes as well as TOEFL tests (interview with NRA Official, 2016). Given that the NRA had hired over one hundred fresh graduates by 2018 and was planning to hire more, this presented an opportunity to train a substantial portion of NRA officials from scratch and to shape a new generation of nuclear safety regulators.

Other training opportunities for officials included higher education within Japan or abroad as well as practical overseas training. In fiscal year 2013, three staff members were sent to graduate school in related fields in Japan, for example at the University of Tokyo. By 2016, at least three staff members had been sent to the IAEA for further training. To gather practical experience abroad, the NRA sent officials to, for example, the US NSC and the Nuclear Energy Agency of the OECD (NRA 2014a, 14; 2015a, 2016a). Critics of Japan's nuclear village will likely point out that the University of Tokyo—and other universities with nuclear engineering departments—were enmeshed in the pre-3.11 system of regulatory capture and question the usefulness of such staff training as a means of securing independent expertise. Despite such criticism, further education at Japanese universities, in combination with in-house training by the HDRC and practical overseas training, is a clear sign of the NRA seeking to raise in-house expertise through diverse means of staff training.

In sum, the first board set out to address those personal connections between regulators on the one hand and administration and industry on the other hand that had contributed to capture, according to the NAIIC report. It did so by partly or completely removing the NRA from standard practices, such as staff rotation between administrative units as well as between the administration and industry. While many of the staff who initially joined the NRA returned, those officials who chose to remain inside the NRA proved determined to foster independent expertise by creating and refining career paths inside the NRA in connection with specialized training.

Institutionalized Transparency, In-House Expertise, Neutrality, and Internationalization

This chapter set out to elucidate which organizational characteristics enabled the NRA to defend its independence and to firmly establish fundamentally different safety governance. To begin with, many aspects were left vague in the NRA Act and had to be specified by the first board, which played a key role.

Controversies over board appointments speak to the difficulty of finding nuclear experts without close ties to the industry. Board members may not come from a position too close to the industry or hold other offices. These rules clearly set the NRA board apart from other agency heads or commission members, which are often appointed by a ministry and hold full-time positions in a related field. The Noda government was confident in its choice of candidates and appointed the first board despite opposition, especially against the first chairman, Tanaka Shun'ichi. The contested Chairman Tanaka was certainly pronuclear, but with the important addition of stressing safety first. He had spent his working life in organizations promoting nuclear power, but he nevertheless turned out to be a stroke of luck for the safety agency. What set him apart was his determination to speak out against flawed nuclear safety practices and to defy critics of the NRA's independent approach. As he was close to retirement age, the new safety agency was an opportunity for him to leave a mark. Trained as a scientist, as NRA Chairman he quickly adapted to the politics of nuclear safety governance and successfully steered the NRA through rough seas. Shōriki Matsutarō is sometimes referred to as the "father of nuclear power," because he was instrumental in establishing the pre-3.11 system with a positive media presence and with the nuclear industry as a partner in nuclear safety governance. Since Tanaka took the lead in establishing transparent, stringent, and independent safety governance, he can be dubbed the "father of improved nuclear safety."[15]

The story of Tanaka and other high-level NRA officials highlights the difficulties of delineating the boundaries of the nuclear village. Referring to a system of close ties between nuclear industry, politicians, and administration, particularly METI and MEXT, it includes close ties between nuclear industry and researchers, an aspect the DPJ selection criteria for NRA board members focused on. All actors in the nuclear village are collective actors, and some of the low expectations regarding the NRA's ability to become independent were clearly rooted in assumptions about collective actors, for example the taking over of staff from the previous safety administration and a pronuclear chairman heading the new agency (Shadrina 2012). Collective actors naturally include individuals who are more supportive and ones who are less supportive of nuclear power. Some may have simply tolerated collusion and lackluster safety standards rather than actively seeking such flawed governance. Unless these individuals speak up, however, there is no knowing where exactly they stand. Furthermore, an event such as the Fukushima nuclear accident has the potential to fundamentally alter one's thinking about nuclear safety. Tanaka, Fuketa, and other high-level NRA officials, who came from collective actors considered to be part of the nuclear village, turned out to be enthusiastic reformers. And, as discussed in chapter 5, even some prominent LDP members, such as Koizumi Junichirō, began speaking out against nuclear power while the LDP overall still supported pronuclear policies. For the sake of simplicity, this book refers to the abovementioned collective actors as members of the nuclear village, but in a descriptive rather than blaming sense, and stresses that the boundaries are fuzzy.

Under Tanaka's leadership, early board decisions expanded the NRA's regulatory powers. The NRA Act grants the NRA sole authority over nuclear regulation. The decision to absorb JNES, an organization of roughly equal size, was tantamount to assuming the power to monitor safety refurbishments. On top of monitoring competencies, the NRA board acquired the power to sanction safety breaches by introducing the backfit system. This put an end to the previous practice of "reciprocal consent," negotiating nuclear policy and safety with the utilities (Samuels 1987). In other areas, however, the NRA only held the power to regulate and monitor. This was the case with nuclear security and safeguards, handling isotopes and radioactive materials, radiation measurement, and handling the damaged Fukushima Daiichi NPP. In the case of emergency response, it lost the power to monitor and sanction to the government. Thus, the NRA's regulatory competencies were strongest in the area of nuclear safety, where it expanded them.

Political influence can find its way through a myriad of channels, such as politicians' influence on the budget process, personal contacts with board members, and the appointment of board members. In terms of financial resources, the NRA was as financially independent as possible within the administration, suffered no budget cuts, and overall seemed to have sufficient financial resources.

Chairman Tanaka's retirement in 2017 was a litmus test for the NRA's susceptibility to political influence. Had he been succeeded by somebody determined to return to pre-3.11 safety governance, it could have changed the NRA's course. Instead, Fuketa Toyoshi became the new chairman. Fuketa was instrumental in enforcing the NRA's thorough safety checks as the commissioner in charge of nuclear safety and, after Tanaka's retirement, he was the last remaining member from the groundbreaking first board. He would determine the NRA's fate until 2022. With Fuketa succeeding Tanaka, who remained an adviser, the NRA clearly showed continuity in leadership determined to keep a distance from politicians, the administration promoting nuclear power, and the nuclear industry.

The NRA achieved a high degree of transparency. Prior to the reforms, the nuclear safety administration was characterized by low transparency, decision-making behind closed doors, and little public participation. The NRA vowed to be transparent in order to win back public trust. Transparency is an important good in democracy by enabling the public to access necessary information and allowing for public scrutiny. NRA guidelines for transparency and administrative documents interlinked and formed a comprehensive information release system not subject to information release requests. Intended to regain public trust in nuclear safety, transparency also kept NRA officials in line and protected the new safety agency against pressure. Following public demand for citizens' participation in the process of restarting nuclear reactors, the NRA added outreach activities. It invited public comments on draft regulation and held meetings in municipalities where nuclear reactor restarts were in progress. Stopping short of full citizen participation, the NRA nonetheless established public accountability with unprecedented transparency and outreach activities.

Securing independent expertise was—and remains—a crucial task, which the NRA embraced. Efforts to that end included creating a career path in nuclear regulation, hiring fresh graduates and midcareer officials, limiting staff rotation with parts of the administration promoting nuclear power, curbing retirement to the industry, and introducing comprehensive in-house training. Building up in-house expertise through these measures enabled the NRA to sustain itself and to make regulatory decisions without relying on industry experts. Furthermore, limiting personnel exchanges with promoters of nuclear power such as METI and the nuclear industry, creating a career path in nuclear regulation, and preventing retirement to the industry effectively curbed incentives for officials to put nuclear energy promotion ahead of safety. Thus, the NRA made good progress toward building independent in-house expertise, but continuing efforts are necessary to sustain and expand current knowledge as well as to adapt to new tasks.

The overarching reform aim to regain domestic and international trust made international actors an important audience. This was also evident in the NRA's internationalization efforts. Along with actively cooperating with international peers, it invited international scrutiny in the form of the IAEA evaluating the NRA's progress towards becoming an independent regulatory agency. With many recognizing the IAEA as an international authority on appropriate safety governance, an IAEA review could lend legitimacy to measures taken by the NRA and potentially push it even further. When initial reform advocates such as LDP reform leader Shiozaki, who insisted that Japan should follow IAEA standards as "an international charter of nuclear regulation" (Y. Shiozaki 2012, 27), later called on the NRA to give more consideration to their political concerns (Amano 2015, 21), international scrutiny in the form of IAEA evaluations lent legitimacy to the NRA's independent approach and helped ward off pressure from pronuclear actors. Furthermore, IAEA recommendations helped legitimize NRA actions that were unpopular with operators, such as more intense and critical safety inspections, putting an end to an inspection system that was more akin to utilities giving regulators a guided tour for which they had to register in advance. As such, active engagement with international peers through the IAEA buttressed NRA independence.

Overall, the very beginning was a crucial time for the NRA's establishment as an independent regulator. At the time, (then designated) board and senior officials worked closely with the outgoing Noda Cabinet to establish a new kind of safety administration organization. Adopting a no-return rule and a no-*amakudari* rule as well as renouncing the journalist club approach to media coverage situated the NRA outside of institutional arrangements that reinforced the previous system of regulatory capture, including officials retiring to the industry they used to oversee, staff exchange within the administration, and the journalist club system. Using the NAIIC report as a set of guidelines for what to avoid, the NRA expanded regulatory and monitoring competencies, in-house expertise, and transparency while curtailing avenues for pronuclear actors' influence by limiting staff exchange with the industry and parts of the administration promoting nuclear power. Subsequently enshrining many of these guidelines in the legal framework as amendments to the NRA Act formalized the vision of an independent, neutral, and transparent organization, which made it considerably harder to undo these new rules later on. In the months before and after the NRA's inauguration in September 2012, the NRA board, top NRA officials, and part of the Noda Cabinet worked together to create high hurdles for those seeking to return to pre-3.11 safety governance and to undermine, or undo, the new institutional arrangement.

This was crucial for post-3.11 reforms to result in a critical juncture in Japan's nuclear safety governance approach and, by extension, nuclear power politics. Explanations of why a window of opportunity turns into a critical juncture stress the notion of agents of change successfully pushing for change. In case of the NRA, some of the policy entrepreneurs who had been involved in drafting the NRA bill for the DPJ government later moved into high-level positions within the NRA. Using the NAIIC findings as guidelines, they proved determined to lend substance to the NRA's legal status as an Article 3 independent regulatory commission. These efforts were supported by a political entrepreneur, the Noda Cabinet, which helped enshrine their vision into the law just before the LDP returned to power. Important outside entrepreneurs during the reforms, international regulatory peers and the IAEA later acted as a kind of watchdog making sure that the NRA was in fact on track to living up to international independent regulation norms in the IAEA evaluations. Hence, during the "crucial founding moments of institutional formation" (Thelen 1999, 387), domestic and international agents of change helped to protect the NRA from pronuclear actors seeking to undermine or undo independent safety governance.

The NRA's ability to act independently in the long run depends on its ability to continuously improve in-house expertise and regulation, to successfully implement its new inspection system, to resist industry attempts to water down safety standards during implementation, and to continue eliciting domestic and international scrutiny. Public attention, however, is a fickle good. If public opinion turns pronuclear again, it will weaken transparency's function as a protective shield, which makes scrutiny by international peers all the more important. The turn toward decommissioning nuclear reactors also presents a challenge as the NRA lacks regulatory powers, especially sanctioning mechanisms, in this area. A promising sign is that the NRA board has shown willingness to continuously improve its regulations. Unless the NRA assumes regulatory capabilities, including sanctioning powers, as it did with nuclear reactor restarts, its future position vis-à-vis the industry will be considerably weaker than it is now. Hence, the NRA's future as an independent regulatory agency hinges on its ability to adapt to new tasks in the same strategic manner it exhibited with nuclear reactor safety as well as on domestic and international scrutiny.

POST-3.11 NUCLEAR SAFETY STANDARDS

Safety standards refer to a set of goals, requirements, and principles the design of a nuclear power plant needs to meet, also called the design basis. The design basis plays an important role as a risk acceptance mechanism, especially for high-risk technologies and facilities. The expectation is that a nuclear power plant that conforms to the design basis poses "an acceptable risk to all stakeholders, including society as a whole" (Wyss 2016, 34).

Prior to 3.11, Japan's safety requirements regulated many technical components of reactor units but allowed electric utilities to largely determine residual risks for society via partly voluntary safety standards, particularly accident countermeasures and emergency response measures. The situation amounted to one of regulatory capture, where the industry, especially electric utilities, prevented safety requirements that experts recommended. As accidents inside and outside of Japan revealed the risks associated with nuclear power, industry and government promoted a safety myth about the absolute safety of nuclear power plants made in Japan instead of strengthening safety measures. The lack of safety governance ultimately led to the devastating nuclear accident (NAIIC 2012).

Strong safety standards require the incorporation of new scientific knowledge from a variety of disciplines. For a country such as Japan, an island nation located atop the Pacific "ring of fire," knowledge from seismology, volcanology, and tsunami research matters for nuclear safety regulation. Japan's previous safety administration suffered from a lack of interdisciplinary communication. An example of this was the "Regulatory Guide for Reviewing Seismic Design of Nuclear Power Reactor Facilities," revised in 2006, which treated tsunamis not as natural hazards in their own right but rather as events accompanying earth-quakes, and it omitted specific safety requirements for tsunamis. Consequently, Shiroyama (2015, 289) concluded that Japan's nuclear safety regulators lacked an "antenna" for developments in fields such as seismology and tsunami research, a conclusion similar to Kikkawa's (2012, 33) call for new safety regulation that "adapts to new scientific knowledge."

The dangers that resulted from the safety myth, partly voluntary safety standards, and a lack of interdisciplinary communication were shockingly illustrated by the March 2011 nuclear accident. Ensuing reforms created a new nuclear safety agency, the Nuclear Regulation Authority (NRA), which stated its approach as follows:

> We must be jealous in protecting NRA's organizational independence and earnest in keeping its operational transparency; we should be careful not to consort with electric utilities and other interest groups; and we will be tireless in our efforts to improve our regulatory measures. (NRA 2014c, 2)

The first two goals, organizational independence and operational transparency, were discussed in chapter 3. With independence from the remainder of the administration, politicians, and the nuclear industry as well as a transparent modus operandi that invites scrutiny by Japanese citizens and international actors, the NRA marked a departure from the previously untransparent and captured regulatory organizations. Whether the new regulators warded off industry influence and sought to continuously improve regulatory requirements is the subject of this chapter.

For the formulation of new safety standards, the NRA board established study teams[1] to discuss different safety aspects. Three study teams developed safety requirements, another three formulated emergency preparedness and response measures, and one designed a safety review system. One after another, study teams presented draft regulations to the public for comments, beginning with draft emergency preparedness and response measures in October 2012 and ending with safety requirements in the spring of 2013. Following public comments, provisions for distributing stable iodine as part of emergency planning were improved. After a period of deliberation and slight amendments, new safety standards for commercial nuclear power plants went into force on July 8, 2013.[2] Once safety standards and a review system were in place, the NRA was ready to conduct safety reviews.

The process of designing new requirements and a safety review process relied on what I termed the council system. Different study groups were in charge of drafting proposals for the NRA board, which adopted them after a few rounds of requesting further details and improvements. Study teams consulted external experts based on neutrality criteria and later opened the drafts for public comment, which the NRA integrated in the finalization process as it saw fit. With the exception of the study group in charge of developing the safety review system, none of them conducted hearings with industry representatives. Thus, the new regulatory standards were developed without influence from the regulated parties, but industry representatives were consulted regarding practical aspects of safety reviews.

The aim of this chapter is to assess how far the establishment of an independent safety regulator with enhanced regulatory competencies has translated into stringent safety regulation, implementation, and monitoring. Different nuclear safety aspects to be considered include basic reactor design, accident countermeasures, and emergency response measures. On top of that, there are nuclear security issues and safeguarding nuclear materials.[3] Out of the three areas of nuclear safety, security, and safeguarding, the NRA holds the most competencies, including sanctioning powers, with regard to nuclear safety. Focusing on nuclear safety, this chapter takes a closer look at new regulatory requirements, their implementation, and the monitoring system. Whether NRA regulatory requirements were in line with IAEA standards is an indicator of the progress made in improving safety standards. Unless these are implemented and monitored thoroughly, there will be room for electric utilities to influence safety levels again. How safety regulations translated from rules on paper into concrete safety measures is indicative of the NRA's stance vis-à-vis electric utilities.

The chapter first lays out the safety standards and then looks at the implementation process, using exemplary safety reviews of nuclear power plants, to determine the NRA's level of stringency in practice. The third section turns to safety inspections and monitoring of measures taken at different power plants. The last section contextualizes the findings in international standards and global regulatory practices.

Binding Safety Standards

Post-3.11 safety requirements were developed to heed lessons from the March 2011 nuclear accident, incorporate IAEA safety standards and guidelines, and overcome Japan's previous safety myth by assuming that severe accidents could occur at any moment (NRA 2013d). As study teams deliberated new safety requirements, the NRA board adopted two requirements it regarded as concrete lessons learnt from the accident, namely a forty-year limit for the operating life span of a nuclear reactor and a "backfit system" (NRA 2013b, 29). Innovations introduced with the "New Regulatory Requirements for Commercial Nuclear Power Plants" can be subsumed under the following umbrella terms: defense-in-depth, severe

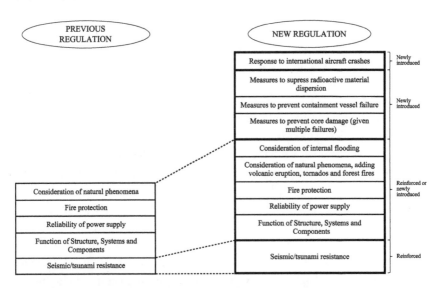

FIGURE 4.1. New safety standards, based on NRA (2013b)

accident countermeasures, backfit system, and a forty-year rule (NRA 2013d). Figure 4.1 provides an overview of the expanded regulations.

The forty-year limit for a reactor's lifespan means operating licenses are only granted for forty years. If a plant operator wishes to run a reactor for longer, it can apply for a one-time extension of twenty years. Such an extension is tied to additional safety investments and a renewed safety check that determines whether the reactor is fit for another twenty years of operation.

The backfit system mandates that reactors be updated in line with new scientific developments rather than at the discretion of electric utilities. Under the backfit system, plant operators have to conform to design basis requirements when they commission the plant and update reactors in operation. Such a system provides regulators with sanctioning powers vis-à-vis electric utilities as they can threaten to revoke a license should the utilities fail to comply. By introducing a backfit system at its own initiative, the NRA assumed sanctioning powers omitted in the NRA Act. In this area, the NRA took decision-making power over the implementation of safety standards from electric utilities.

The reactor design basis, specifying which predefined threats a nuclear power plant needs to be able to withstand, was strengthened. New standards require more measures against volcanic eruptions. For this purpose, volcanoes within a 160-kilometer radius are to be surveyed to assess the risk they pose. Furthermore, key facilities of a nuclear power plant, such as a nuclear reactor, may not be situated above an active fault line. An earthquake fault line is considered active when geological layers that are approximately 120,000 to 130,000 years old show signs of displacement or deformation. Mandatory seismic ground motion assessment is conducted to determine the strongest earthquake a nuclear power plant needs to be able to withstand. Another newly introduced requirement is a tsunami protection wall, which needs to exceed the largest tsunami ever recorded in the area. These strengthened design basis requirements are meant to prevent the nuclear power plant from damage.

In case the reactor design basis proves insufficient, there are severe accident countermeasures, which are related to the idea of defense-in-depth, referring to measures against severe accidents with more than one component failure. Accident countermeasures address natural phenomena—including earthquakes, tsunamis, volcanic eruptions, tornadoes, and forest fires—as well as other events, such as an intentional airplane crash by terrorists, fire inside a reactor, internal flooding, and power supply failure. This is the second area where the NRA attained power over the implementation of safety standards.

Accident countermeasures require waterproof doors, fireproof cables, redundant onsite power sources, and switchboards in diverse locations. To increase the reliability of offsite power sources, safety standards mandate connecting at least two power substations with at least two transmission lines and mobile power units on a hill nearby to ensure steady electricity supply even if some equipment suffers damage. To strengthen cooling capacities for the spent fuel pool and the reactor core, mobile water injection systems, and permanently installed water injection systems are required. To prevent a hydrogen explosion at boiling water reactors, such as the Fukushima Daiichi reactors, measures to reduce pressure within a reactor consist of installing a filtered venting system to let out hydrogen if needed. Last but not least, a radiation-protected earthquake-resistant control room, located at least one hundred meters away from other facilities, is required to allow remote-control operation of the plant's cooling systems, for example in case of an airplane crash.

New safety requirements fall into two categories, relative and absolute. Relative standards' implementation depends on prior risk assessments, for example, of the risk posed by volcanoes in the plant's vicinity. Similarly, earthquake countermeasures are based on assessing potential earthquakes and the surface structure to determine the maximum amount of ground motion. Measures to prevent a tsunami flooding the plant are to be based on historical tsunami records. The implementation of relative safety requirements essentially hinges on risk assessment of volcanoes, earthquakes, and tsunamis, which opens up room for contestation and has been somewhat controversial.

New safety requirements aimed to bring Japan's nuclear safety regulation in line with IAEA standards. The chairman of the Independent Nuclear Accident Investigation Committee (NAIIC), which the NRA took many cues from, lambasted the NRA safety standards in a 2015 lecture: "I have personally heard the accounts of many internationally recognized experts who have stated that Japanese nuclear safety requirements remain inferior to the IAEA standards" (Kurokawa and Ninomiya 2018, 48). After the IAEA conducted a review of the NRA as part of its Integrated Regulatory Review Service (IRRS), it heralded the

> NRA's prompt and effective incorporation of lessons learnt from the Fukushima Daiichi accident in the areas of natural hazards, severe accident management, emergency preparedness and safety upgrades of existing facilities. (IAEA 2016c, 2)

At the same time, the IAEA criticized the NRA for making important aspects voluntary rather than mandatory for the operators, including periodic reporting on probabilistic safety assessment and deterministic safety analysis, natural hazards analysis, and equipment condition (IAEA 2016b). A generally tough critic of nuclear power, Greenpeace issued a report that found fault with the NRA's risk and hazard assessments more generally. Probabilistic risk analysis, which has become a benchmark for safety evaluations worldwide, was not a requirement under the NRA's safety regulation. Rather, the NRA required probabilistic hazard analysis, which is similar but less sophisticated with regard to potential risk scenarios considered (Vande Putte, Ulrich, and Burnie 2015). Hence, new safety requirements led to significant improvements, but proponents and opponents of nuclear power alike criticized substandard risk and hazard assessment regulation.

Expanded Emergency Response Planning

Emergency response in case of an accident is a cooperative endeavor that involves different actors. The legal framework—the Basic Act on Disaster Control Measures and the Nuclear Emergency Act—requires electric utilities to cooperate with local governments. Utilities are tasked with preparing an emergency action plan for what happens on site while responsibility for preparing evacuation plans for the areas around the plant lies with local governments. The national government's responsibility is to ensure cooperation on the local level. In addition, the national government's task is to coordinate relevant units, such as police, firefighters, and Self-Defense Forces, to be deployed if necessary (NRA 2013a, 76).

An important emergency response stipulation specifies the radius around a nuclear power plant for which crisis response and evacuation plans need to be prepared. The new guideline distinguishes an "urgent protective action planning zone" (UPZ) and the "precautionary action zone" (PAZ), which describe a five-kilometer and thirty-kilometer radius, respectively. In case of an accident, residents within the UPZ are required to evacuate immediately. In the PAZ, evacuation of residents depends on the circumstances, such as the release of radioactive materials and forecasts about their spread based on the System for Prediction of Environment Emergency Dose Information (SPEEDI) and weather forecasts (NRA 2013b). For both zones, local governments are required to draw up emergency response plans in advance. Mandatory emergency response plans thus require cooperation between at least one electric utility and one local government. In some cases, an electric utility and multiple local governments are required to work together due to an expanded UPZ crossing local district boundaries. With the new UPZ going beyond previous requirements, NRA emergency response guidelines created the need to cooperate more broadly than before.

However, the NRA was not in charge of implementing and monitoring emergency planning. Based a vague phrase in the NRA Act, which put the NRA in charge of technical aspects and the prime minister in charge of operational aspects, such as coordinating relevant ministries and agencies, the Nuclear Emergency Preparedness Commission (NEPC) at the Cabinet Office monitored the implementation of emergency response guidelines. In case of violations, the NEPC held sanctioning competencies, such as withholding consent to a proposed evacuation plan, but practically rubberstamped utilities' emergency response plans.

Safety Reviews

Beginning in July 2013, electric utilities were invited to put nuclear reactors forward for safety reviews. Within one month, the NRA received twelve requests, and another fourteen followed within roughly two years. Electric utilities appeared eager to seek safety checks in order to bring nuclear reactors back into operation. In light of the high costs of replacing missing nuclear power capacities with fossil fuel–based thermal power generation, it seemed to be a step toward many more requests. Defying such expectations, however, utilities sought only one further safety review over the next several years, bringing the number of reactors put forward for review up to twenty-seven by 2022.

The NRA essentially established a two-step procedure for safety checks that first conducts document-based examinations and later on-site inspections. The process begins with an electric utility submitting documents, including construction plans, design basis descriptions, and documentation about proposed construction work. Document-based examinations cover three areas: permission for changes in the reactor installation license, approval of plans for construction works, and approval of implemented safety measures (NRA 2013b). Due to some additional measures, such as tsunami protection walls, constituting gigantic building projects that need to be finished before the last step of the integrated document review, a few years can easily pass between the approval of plans and the approval of changes in operational safety programs. After successful pre-start-up inspections, the utility receives permission to start up the nuclear reactor. Once the reactor passes the post-start-up inspection, the utility may begin generating electricity for commercial purposes again.

For the document-based review, the NRA established a council labelled "review meeting,"[4] including at least one board member, the commissioner in charge of nuclear safety, and about ten NRA Secretariat officials. First, utility representatives presented previously submitted documents proposing measures to meet reactor safety requirements. Following the presentation, review meeting members asked questions about provided materials and, depending on the nature of the question, responses varied. In response to quick clarification questions, utility representatives had the chance to make verbal clarifications during the meeting. In case of critical comments about intended construction work, responses ranged from a debate about appropriate safety measures to an order to rework the materials and resubmit them for the next meeting. Final decisions were made by the NRA board rather than the review meeting. Throughout the process, review meeting members regularly reported back to the NRA board. With recordings and minutes of review meetings and board meetings available online on YouTube and the NRA website, transparency in the review process amounted to publicly accountable communication between utilities and regulator.

The review meeting was the NRA council with the highest meeting frequency. Based on a meeting list on the NRA website, it convened 113 times in fiscal year 2014, 132 times in fiscal year 2015, 112 times in fiscal year 2016, 101 times in fiscal year 2017, 136 times in fiscal year 2018, and 157 times in fiscal year 2019.[5] Judging from the length of video recordings on YouTube, one review meeting lasted between 90 and 360 minutes. With an average length of three-and-a-half hours per meeting plus preparations and reporting to the NRA board, there appeared to be a limit to the number of review meetings the safety agency was organizationally capable of.

Sendai

The first safety review that took place was for the Sendai nuclear power plant, operated by Kyushu Electric. This plant had been refurbished to meet new standards for earthquake and tsunami countermeasures published in 2009, which promised a swift safety review. NRA Commissioner Shimazaki, however, raised the stakes based on a new, and as he argued improved, way of calculating earthquake risks. This became an issue for Kyushu Electric's effort to restart Sendai. The requirement to take additional steps delayed the swift restart the utility had hoped for, prompting a Kyushu Electric executive to complain that "Shimazaki made us suffer" (Asahi Shimbun 2014a). Risk assessments regarding earthquake and volcanic risks faced by the Sendai plant were contested too. In the seismological assessment, Kyushu Electric excluded a fault line off the coast, despite the displacement it causes extending to near Sendai. In response to criticism, Kyushu Electric insisted that it had taken all relevant sources of an earthquake into account and that its report was correct. As part of the safety review process, the NRA accepted the report. It included verified earthquake sources, but not all potential sources, a decision criticized by some Japanese seismologists (Tateishi 2015). Regarding the risk of a volcanic eruption, the NRA followed Kyushu Electric's assumption about negligible risks, despite the island of Kyushu being known for its volcanic activity and the Japan Meteorological Agency warning of a potential eruption (Japan Times 2014b). Consequently, the NRA enforced only limited measures against the potential risk that ash piling up on rooftops of nuclear power plant facilities might threaten building stability. While strict on measures against verified earthquake sources, the NRA was more lenient with risk assessments regarding earthquakes and volcanoes.

Monju

The case of Japan's fast breeder reactor Monju revealed the regulatory agency's strict side. As part of an effort to complete a nuclear fuel cycle, Japan's first and only fast breeder reactor, operated by the Japan Atomic Energy Agency (JAEA), started operations in 1994. Monju, however, was rarely in operation and was problem ridden from the start. In 2013, the NRA took issue with how JAEA maintained the fast breeder reactor. After irregularities surfaced during on-site safety inspections, the NRA ordered the JAEA to revise Monju's operational safety program. The JAEA, however, submitted a report on the matter before it had completed revisions ordered by the NRA. More irregularities were discovered during another on-site inspection (NRA 2014a). With problems persisting, the NRA issued its first recommendation (*kankoku*) in November 2015, which concluded that the "JAEA does not have the capacity to operate Monju safely." The recommendation indicated that a failure to find a new operator would lead to the NRA decommissioning the facility (NRA Commission 2015d). In itself, the recommendation was not binding, but the NRA lent substance to it by threatening to otherwise shut down the facility at the heart of Japan's nuclear fuel cycle ambitions. With no suitable operator in sight, the government made the decision to decommission Monju in 2017. The NRA proved its willingness to shut down an essential nuclear fuel cycle facility over apparent safety mismanagement.

Antiterrorism Measures

For antiterrorism measures, the NRA granted plant operators a five-year grace period. Utilities were required to complete them within five years after the NRA approved construction plans, rather than before the pre-start-up inspections. This applied to the construction of radiation- and earthquake-resistant remote-control centers with multiple power-supply sources to maintain functions in case of power outage and located far away enough to remain operational in case of an airplane crash. In early 2019, none of Japan's nuclear power facilities were in line to meet the deadline. Utilities with nuclear power plants in operation, Kyushu Electric, Kansai Electric Power Company (KEPCO), and Shikoku Electric, asked for the deadline to be extended. During a board meeting in April 2019, the NRA turned down the request and, instead, instructed the utilities to halt operations once they passed the deadline (NRA 2019b). Approved first, the Sendai plant was the first to approach the end of the five-year grace period. As it failed to complete construction work for a remote-control center, Sendai Nos. 1 and 2 went into shutdown in March and May 2020, and Takahama Nos. 3 and 4 in August and October 2020, respectively. They remained halted until construction was completed and approved by the NRA. In late 2020, Kyushu Electric completed the emergency control rooms for Sendai Nos. 1 and 2 and was allowed to restart both reactors, while KEPCO's Takahama Nos. 3 and 4 remained idle as of early 2022. The NRA not budging to utilities' requests to lengthen the grace period poses yet another hurdle for nuclear reactor restarts, unless utilities considerably speed up construction.

Earthquake Fault Line Investigations

In addition to standard safety reviews, the new safety agency also conducted investigations of geographical fracture zones or fault lines to determine whether they are able to produce an earthquake. Fault line investigations differ from the standard safety review in that they are not contingent on requests by utilities. The NRA safety requirements clearly state that reactor buildings and other key facilities may not be located atop an active fault line. Initially, the NRA board decided to conduct fracture zone investigations at the same six nuclear power plants the NRA's predecessor began to investigate after earthquake safety guidelines were revised in 2006, namely Higashidōri, Shika, Mihama, Ōi, Tsuruga, and the fast breeder reactor Monju.

For each investigation, the NRA set up an expert meeting that worked independently of the safety review process. External expert selection followed the NRA's Requirements for Ensuring Transparency and Neutrality, which include a disclosure of information on the expert's relationship with the electric utility under investigation. Eligible experts have not been involved in earlier examinations of the fracture zone, have not held executive positions within the electric utility, and have not received more than ¥500,000 in renumeration yearly from a utility (NRA 2013b, 59). The experts, recommended by, for example, the Japan Society for Active Fault Study or the Seismological Society of Japan, did not exchange opinions with the respective utility. Rather than to rely on information provided by utilities, they collected their own data. In this case, data collection included a strong element of on-the-ground activity, such as conducting on-site excavations of fault lines.

In case of the Tsuruga No. 1 reactor in Fukui Prefecture, the operator, the Japan Atomic Power Company (JAPC), claimed the fault line was inactive and applied for a restart permit. NRA and external experts conducted excavations around the fault line to collect their own geological data, based on which the expert meeting concluded the fault line to be active (NRA Commission 2014). This finding did not lead to the reactor being designated for decommissioning, which is a decision the utility has to make. Rather, the discovery of an active fault line means the NRA will not grant an operating license should the utility seek a restart. With Tsuruga No. 1 unable to receive a restart permit, JAPC permanently shut the reactor in April 2015 and began preparations for decommissioning.

Regarding the Shika power plant, the expert meeting concluded that the fault running beneath reactor no. 1 was likely to be active. They found two more faults running beneath key facilities for reactor no. 2 (NRA Commission 2016). By late 2019, the operator, Hokuriku Electric Power, had neither sought a restart of reactor no. 1 nor shut it down permanently, simply leaving it idle. The safety review of reactor no. 2 was complicated by the fact that key facilities for reactor no. 2 might have to be relocated. This has not dissuaded the operator from seeking a restart permit, a process ongoing as of September 2022.

Fault line investigations at Higashidōri proved difficult. While the expert meeting agreed on the two main faults running underneath the site being active, uncertainties remained about whether the fault running beneath reactor no. 1 and important facilities could in fact cause an earthquake (NRA Commission 2015c). Similar to the case of Shika, the operator of Higashidōri applied for a safety review in June 2014, also ongoing as of September 2022.

For the other sites, Ōi, Monju, and Mihama, the expert meeting concluded that the fault line was inactive. The NRA further investigated Takahama, Ikata, and Sendai to determine whether fault lines underneath were capable of causing an earthquake. These additional investigations were conducted by the NRA instead of external experts and concluded that the fault lines were inactive (Ishiwatari 2016).

Safety Review Progress

A decade after 3.11, and nine years after it began safety checks in 2013, the NRA had completed seventeen out of twenty-seven requested reviews. Time-consuming and thorough safety reviews defied political pressure for swiftly bringing nuclear power plants back into operation, such as Prime Minister Abe's call to finish safety checks of all available commercial reactors by 2016 (Abe Shinzo 2013). The lengthy review process posed a challenge for utilities and the NRA as reactors continued to age and to approach the forty-year limit while safety reviews were underway.

To avoid nuclear reactors going into permanent shutdown by default, which could be seen as a sign of the new regulators being unequipped to handle the task, the NRA made exceptions to the forty-year rule. As Takahama Nos. 1 and 2, Tokai No. 2, and Mihama No. 2 were nearing the end of their forty-year life spans, the NRA preapproved their operating licenses beyond the forty-year limit. Even with a preapproved lifetime extension, there will be final inspections once the refurbishment process is complete. If the final inspection follows the same standards as those conducted at other reactors, the preapproval decision resembles a technical difference. If, however, final inspections of construction changes, scheduled to be completed by late 2022 and mid-2023, are conducted less thoroughly, because the license had been preapproved, it would signify a problematic rollback of the forty-year rule.

There were concerns that the review meeting process opens up room for the utilities to bargain about the easiest and cheapest way to implement safety requirements. In the review meetings, regulatory agency and regulated industry meet to talk about how regulations are to be implemented. This setup garnered criticism from Japan's oldest antinuclear civil society organization, the Citizen Nuclear Information Center (CNIC). CNIC head Ban Hideyuki scorched review meetings as an opportunity for utilities to haggle over necessary safety measures and to sound out the cheapest and easiest way to meet new requirements (interview with Ban Hideyuki, 2015). At the same time, rising safety costs indicate the magnitude of refurbishments enforced by the NRA. Japanese newspapers have regularly asked utilities to estimate the costs of investments for updating

the reactors undergoing safety checks. In the fall of 2013, utilities estimated total costs of ¥1.7 trillion (roughly US $17 billion). By April 2015, that figure had
risen to ¥2.4 trillion (Tokyo Shimbun 2015a), and it reached ¥5 trillion (a little
less than US$50 billion) by July 2019 (Asahi Shimbun 2019a). The numbers show
that the investment costs utilities were planning for have skyrocketed since 2013.
Presupposing Ban's assessment of review meetings as being a form of haggling
over the costs of meeting safety requirements, utilities apparently underestimated
the NRA as a negotiation partner.

Inspection System and Safety Culture

Effective safety standards require thorough monitoring of regulated facilities and
a safety-minded operating culture. The inspection system for nuclear power
plants was initially transferred without alteration from the NRA's predecessors.
After its review in 2016, the IAEA made a number of recommendations regarding inspections and the safety culture at nuclear power plants. One recommendation was to amend the legal framework to allow for more effective inspections
(IAEA 2016c, 2). Soon after the IAEA review mission gave its recommendations
in April 2016, the NRA board resolved to overhaul the system with greater freedom for inspectors to select inspection items and issues in mind. Regarding the
traditional inspection system, which narrowly set the timing and scope of inspections, an official working on the new inspection system described it as "a
kind of inspection, but not equivalent to looking at real safety." A legal reform
was necessary, because existing laws prevented an inspection of all safety-related
aspects at the same time (interview with NRA Official, 2018).

A Study Team on Revising the Inspection System drafted an interim report
and, after public comments, submitted it to the NRA's Standing Council on Reactor Safety and Reactor Fuel Safety Examination for further input. The NRA board
asked the standing council, which was established to guarantee outside input and
to assess the NRA's performance, to evaluate its response to IAEA recommendations, and to provide advice. Between July 2016 and January 2017, NRA Secretariat officials reported back to the standing council about the state of deliberations
and decisions. From this process emerged a bill to revise the Act on Regulations
of Nuclear Source Material, Nuclear Fuel Material, and Reactors in order to improve the inspection system, which the NRA board approved in February 2017
and submitted to the Diet (NRA 2018). The bill to alter the inspection system was
approved by the Diet and enacted in 2020.

According to the NRA official involved in revising the inspection system, the
aim of the legal reform was to combine narrowly defined inspection programs for

installing new facilities or periodic safety tests, among others, into one comprehensive inspection system that allowed more freedom for inspectors. The new inspection program widened the area of inspection and gave the NRA more freedom to subject plant operators to penalties, mainly by increasing inspections in reaction to unsatisfactory results. In 2018, the NRA was in the process of overhauling inspection manuals. With about 80 percent of the new manuals done, it embarked on a test run to identify potential challenges and obstacles. The new manuals follow a different logic than previous checklists and employ a coloring scheme. Inspectors are asked to look at different safety aspects with regard to their safety significance and implementation. Only minor safety aspects that are flawless may be colored green by the inspectors. Other issues are to be reported back to the NRA, which takes over deliberations and decision-making. The rollout of the new inspection system was considered to be a challenge because it required a change in mindset, and it looked set to undergo a gradual process of improving the system beyond 2020 (interview with NRA Official, 2018). The new inspection system simultaneously increased and limited inspectors' competencies. They were given more freedom to move around the plant and inspect more widely as they saw fit. At the same time, inspectors had less decision-making power and needed to report back to the NRA for final decisions on anything but minor issues.

An inspection system that relies on inspectors autonomously identifying safety issues requires competent inspection staff. Upon the NRA's request for additional advice, the IAEA review mission recommended a tailored training program for inspectors, international input on inspector training, and holistic training including communication skills and the development of a questioning attitude (IAEA 2016b, 108). Revising its overall staff training system, the NRA established tailored programs for inspectors, nuclear safety reviewers, safeguards inspectors, emergency preparedness officials, and radiation regulators. Seeking international input, the NRA sent a study team to the US Nuclear Regulatory Commission (NRC) in July 2016 to learn about its inspector training system and the required qualifications (NRA 2018, 60). Following that, the NRA invited an inspector trainer from the NRC to help develop the tailored accreditation program for inspectors and to hold seminars on plant inspection best practices and communication with operators and workers on site (interview with NRA Official, 2018). Unlike other NRA training programs, inspectors needed to pass a final exam to gain an inspector license. An increasing focus on inspections was also reflected in a July 2017 secretariat restructuring, which clearly separated four licensing and three oversight divisions within the nuclear regulation departments (NRA 2019a, 11). Improving the skills of safety inspectors was one aspect Chairman Fuketa stressed at his first press conference as head of the NRA (NRAJapan 2017).

With a new legal framework, overhauled manuals, a tailored training program, and an adjusted organizational structure, the NRA looked set to introduce the new inspection system. In early 2020, an IAEA follow-up mission to its 2016 review revisited areas for which it had previously made recommendations. It lauded the establishment of an inspection program that allows unannounced inspections as an important achievement. The review team leader, Ramzi Jammal, stressed the need for the NRA to continue on the path of strengthening safety, including an effective implementation of the new inspection system (IAEA 2020a). Changes already implemented and efforts underway by mid-2022 were promising and revealed an apparent dedication to continuous improvement, but they had yet to prove themselves in practice.

One aspect not directly subject to regulation but nevertheless a concern to the NRA is the safety culture of plant operators and workers on site. Since around-the-clock monitoring by regulators is not an option, nuclear safety depends to a large extent on the attitude of plant operators and workers. In Japan, a number of instances have revealed a questionable safety culture. Prior to 3.11, TEPCO drew negative attention for falsifying reports about safety inspections conducted at nuclear power plants. As mentioned above, Monju's operator JAEA had its operating license revoked as a result of its inadequate safety culture. A former member of the Nuclear Safety Commission stressed the importance of but also the difficulties involved in promoting a safety culture as it cannot be prescribed (interview with Former NSC Member, 2016). In a similar vein, the IAEA called on the NRA to strengthen its promotion of a safety culture at nuclear power plants and to foster a questioning attitude (IAEA 2016c, 2).

The NRA turned its attention to the safety culture at nuclear power plants once new safety standards were in place and safety reviews underway. In October 2014, the NRA began exchanging views with plant managers and electric utilities to promote a safety culture in Japan and to identify necessary safety improvements. During his first speech as chairman, Fuketa Toyoshi stressed the need to improve operators' safety culture, to increase operators' ability to undertake safety checks, and to make them take greater responsibility for the safety of their power plant. In a nutshell, he urged operators to develop a "sense of my plant, my responsibility" (*mai-puranto-ishiki*) (NRAJapan 2017). At a February 2017 board meeting, the NRA confirmed that it would continue exchanging views with plant managers and electric utility executives about once a month to improve the safety culture (NRA 2018, 3). To address the safety culture challenge, the NRA developed a two-sided approach to fostering a better safety culture at Japan's nuclear power plants by talking to plant managers and electric utilities to gain their understanding and by punishing those who fail to comply.

Conventional rather than Precautionary Risk Regulation

This chapter assessed whether the creation of an independent safety regulator translated into stringent safety regulation, implementation, and monitoring. In doing so, it considered different nuclear safety aspects, such as basic reactor safety, accident countermeasures, and emergency response measures, as well as the inspection system and safety culture.

New safety standards were considerably broader in scope and more stringent than before. In a major change, accident countermeasures became subject to state regulation instead of implemented at electric utilities' discretion. Furthermore, with the backfit system the NRA gained leverage over electric utilities via threatening to revoke an operating license should they fail to comply. It gave the new safety agency the means to prevent the recurrence of situations such as those in the 1990s and 2000s, when electric utilities ignored the nuclear safety regulators' recommendations for stricter safety measures. With binding safety regulation and a safety agency with the capacity to enforce backfitting, Japan's nuclear safety regulation after 3.11 improved considerably.

Many new requirements addressed failures that led to the March 2011 nuclear accident and the flaws it exposed in Japan's emergency response system. Beginning with emergency response, thorough emergency planning and the immediate use of SPEEDI became mandatory for a PAZ of thirty kilometers around a nuclear power plant. The scope of safety regulations varied from area to area, with a clear focus on lessons learnt from the Fukushima Daiichi nuclear accident. Absent were new threats such as a cyberattack or other IT-related safety risks. Hence, the new safety requirements were geared toward avoiding an accident such as the Fukushima nuclear accident and left room for improvement in other areas.

The nature of risk assessments garnered the NRA criticism from both proponents and opponents of nuclear power. Considering risk assessment practices worldwide, there are two distinct approaches to scientific uncertainty and assumptions about risks, the traditional and the precautionary. Since the 1990s, the precautionary principle has often been associated with stringent EU consumer and environmental regulation whereas the United States represents the traditional approach (Vogel 2012). The precautionary principle, a term popularized by environmentalists and environmental protection treaties, is the

> presumption that in situations where there are threats of serious or irreversible damage, lack of full scientific certainty should not be used as a reason for postponing cost-effective measures to prevent environmental degradation. (Sampson 2002, 60)

The precautionary principle shifts the burden of proof in risk assessment on proponents to show that potential risks will be of smaller scale than opponents envisioned. In cases where the magnitude of potential risks is considered too great, a product or technology is presumed harmful until proven safe. In contrast, the traditional risk regulation approach assumes the same technology to be safe until proven harmful. In cases where science cannot fully establish the nature and magnitude of risks, proponents of a product or technology with the potential to greatly benefit the environment or human health push for an implementation with due caution, often resulting in a battle over the precise meaning of "due caution" (Durant 2004). Nuclear power is a good example of a technology that inspires hopes of great benefits but also instills fears of high risk, with science unable to resolve uncertainties about the risks involved.

Efforts to restart Japan's nuclear power plants are fraught with fights over due caution. These play out in cases of scientific uncertainty, where risk regulation becomes a matter of interpretation and assumptions about worst-case scenarios the plant could potentially face. In the case of Sendai, and other safety reviews for that matter, the NRA accepted utilities' risk assessments, which only included proven sources of risk and assumed that all other potential hazard sources were negligible. In doing so, the NRA fell in line with industry risk interpretation in areas of scientific uncertainty instead of taking a more precautionary approach. Accepting utilities' risk assessments despite warning voices garnered the NRA criticism from the CNIC (interview with Ban Hideyuki, 2015) and Greenpeace (Vande Putte, Ulrich, and Burnie 2015) for failing to thoroughly implement safety standards. Citizen groups opposed to nuclear power blasted NRA decisions on risk assessment for omitting potential sources of risk and called for a precautionary approach, and the IAEA criticized certain risk assessments for being voluntary rather than mandatory (IAEA 2016b). As chapter 5 shows, opponents of nuclear power filed class-action lawsuits that question the risk and hazard analyses conducted by the NRA and, thus, its conventional risk regulation approach.

The NRA rejected the precautionary principle, but it implemented rigid regulation within the bounds of conventional risk regulation. Looking at the NRA's regulatory decisions, Monju is an example of taking a tough stance on a lack of safety culture and faulty reporting, Tsuruga exemplifies a strict stance on verifiable earthquake fault lines, and the shutdown of two Sendai units highlights an unyielding position in case of utilities failing to implement safety measures in time.

The conventional risk regulation approach has a clear advantage for the NRA in terms of defending its independence. Safety issues have long been at the heart of the struggle over nuclear power plants. Pre-3.11 safety debates about earthquake fault lines became a political issue. Over time, improvements in seismology led to the discovery of more fault lines with the potential to shut down nuclear power plants, which made the regulatory response essentially political in nature. As Scalise (2015) has argued based on restart debates surrounding three reactors between 2011 and 2013, two under the NRA's predecessor and one handled by the NRA, scientific agreement enhances scientific authority. Vice versa, a lack of scientific consensus opens up room for political debate and maneuvering as well as public mobilization. With nuclear safety, and earthquake resilience in particular, as a politicized topic, the NRA erred on the side of caution with regard to the safety agency's political positioning. By rejecting the precautionary approach of including potential but ultimately unverifiable risks, which civil society organizations and some scientists called for, the NRA upheld its proclaimed neutrality by apparently making apolitical judgements based on scientific certainty.

In a nutshell, the new safety agency formulated and enforced broadly applicable and considerably more stringent safety standards. During a press statement on the IAEA safety review results, Chairman Fuketa pledged that the "NRA will never become complacent regarding nuclear safety" (IAEA 2020a). Regulators showed their determination to heed this pledge by overhauling the inspection system. The NRA's responsiveness to IAEA recommendations has been much greater than before, which is a testament to a self-critical attitude and openness to continuous improvement. In contrast, the NRA had yet to apply the same rigor to new threats and risk assessments. The hesitation to adopt a more precautionary approach that considers uncertain risks presented a way to shield the NRA from additional pressure amid an ideologized debate about nuclear power but put it below the international standards it aspired to catch up with.

5

THE FISSURED "NUCLEAR VILLAGE"

Every crisis offers an opportunity for change. After 3.11, many scholars asked whether Japan would phase out nuclear power altogether. The nuclear phaseout debate (discussed in chapter 2) took place amid large antinuclear demonstrations but resulted in no more than a vague DJP government decision to reduce Japan's dependence on nuclear power. One reason was severe nuclear industry opposition to a zero nuclear power policy. Soon after, the pronuclear LDP won the 2012 general elections in a landslide victory, public demonstrations subsided, and pronuclear actors reasserted the necessity of nuclear power in the public discourse, all of which were taken as clear signs of continuity in Japan's nuclear energy policy (Scalise 2014; Kingston 2014; Cotton 2014). In contrast to the aborted phaseout debate, nuclear safety reforms came to fruition. In order to regain domestic and international trust lost after 3.11, a triparty compromise between the DPJ, the LDP, and Kōmeitō resulted in the creation of the Nuclear Regulation Authority (NRA) as an independent nuclear safety regulator. Hence, the Abe government signaled a return to nuclear power, albeit with an overhauled nuclear safety administration.

Nuclear power has long been an important instrument to raise Japan's energy self-sufficiency rate, which reached a low point after 3.11 (Vivoda 2014), and to lower greenhouse gas emissions, which peaked after 3.11, in order to meet Japan's commitments under the United Nations climate regime (Kameyama 2019). In addition, reinvigorating the nuclear industry was so central to Abe's three-pronged economic policies to reignite growth, called "Abenomics," that Kingston (2016) dubbed it Abenomics' "fourth arrow" and Incerti and Lipscy

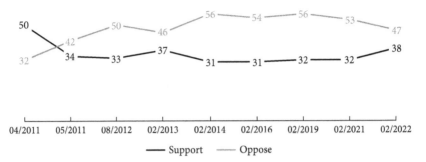

FIGURE 5.1. Opinion Poll conducted by the *Asahi Shimbun* on the use of nuclear power, shown in percent, created by the author based on data from Shibata and Tomokiyo (2014) and Asahi Shimbun (2016b, 2019b, 2022). Question asked until 2014: "Do you support or oppose the use of nuclear power?" Question asked since 2016: "Do you support or oppose the restart of idled facilities?"

(2018) created the moniker "Abenergynomics." Even though the pronuclear LDP won the 2012 elections and mass protests subsided, there was a lack of public support for nuclear power. As shown in figure 5.1, from early 2011 to 2022, the ratio of citizens supporting and opposing nuclear power basically reversed. In 2019, a majority of 56 percent opposed restarting nuclear power plants and only 32 percent supported such a move. By 2022, however, support and opposition were more balanced again. The government pursued nuclear power against public opposition to achieve long-standing energy security and climate policy aims as well as Abe's signature policy of invigorating economic growth.

However, almost a decade after 3.11, the future of nuclear power remained up in the air. Operational nuclear power plants accounted for only 6 percent of electricity generation in 2018–19, according to IAEA data. Apparent support by powerful political actors has not translated into swift policy implementation, raising questions about the nature of nuclear politics after 3.11.

The nature of politics depends on the institutions that structure political interaction. Formal and informal rules (institutions) structure the political interaction of collective actors (organizations). In other words, institutions, and the organizations they are embedded in, define actors' access to political agenda setting and their decision-making power (Thelen 1999). The process of designing new institutions produces new winners and losers by constraining and enabling different actors than before. Actors design institutions to be change resistant in order to constrain their political successors or, particularly in times of crisis, to make their commitment to an institutional solution more credible (Pierson 2000b). During the time it takes for new institutions to take root, it is crucial for initial winners to defend the arrangement against attempts to undermine and reshape it in favor of disadvantaged actors. If they can sustain the new institutional design, it puts the institutional arrangements, and with it the nature of politics, on a different path.

Understanding Japan's surprisingly sluggish return to nuclear power warrants a closer look at the NRA, Japan's first independent and transparent nuclear safety agency and beneficiary of post-3.11 nuclear safety administration reforms. The NRA addressed many of the flaws that had previously besieged nuclear safety governance (discussed in chapter 3). Its achievements included separating nuclear power promotion and regulation, raising in-house expertise, strengthening public accountability via far-reaching transparency, and imposing binding safety regulation on the nuclear industry, particularly electric utilities. In the process, the NRA limited staff exchange with the industry and parts of the administration promoting nuclear power, which curtailed avenues for pronuclear actors in the industry and administration to influence safety regulation. Seeking to "restore public trust, in Japan and abroad, in the nation's nuclear regulatory organization" (NRA 2013b, 3–4), the NRA invited the Japanese public and international nuclear safety experts to scrutinize its work. The creation of the NRA marked a turning point in Japan's nuclear safety administration with more independent, transparent, and stringent safety governance.

This chapter highlights the NRA's interactions with other actors regularly involved in nuclear politics. First, this chapter turns to relations between the NRA and pronuclear actors seeking a swift return to nuclear power, those that are sometimes labelled the "nuclear village." This section traces how the NRA stood its ground against attempts to undermine its position of power and, in the process, forced the Abe government, the LDP, and government ministries to adapt their strategies. The second part turns to the response of the nuclear industry, especially electric utilities, to the new safety governance environment. Following that, this chapter turns to antinuclear groups as the ones seeking access to and challenging the NRA's regulatory decisions. The findings document how independent and transparent safety governance redistributed power in Japan's nuclear politics, especially with regard to policy implementation.

Political Support and Pressure

The Abe government adopted the 2014 Strategic Energy Plan (SEP, *Enerugī-kihon-keikaku*) in April 2014, which built on the 2010 Strategic Energy Plan in many ways, but also included new elements. Recurring elements were the "3Es", the goals associated with energy policy, namely economic efficiency, energy security, and environmental friendliness. The 2014 Strategic Energy Plan resolved to promote (*suishin*) nuclear power as a "baseload power source" (*jūyōna-bēsurōdo-dengen*) based on the premise of "safety first" (*anzensei wo subeteni yūsen*) and the aim to "reduce dependence on nuclear power as much as possible" (*kanōna kagiri teigen*). (METI 2014). With these disparate aims in the 2014 Strategic Energy Plan, "the traditional lines of Japanese nuclear policy had neither been significantly changed nor clearly reaffirmed." (Hymans 2015, 119) After an additional year of deliberations among representatives from the LDP, Kōmeitō, METI, and the MOE in the Advisory Committee for Energy Policy (*Sōgō-shigen-enerugi-chōsakai*), they agreed on numerical targets for each electricity source. The 2015 Long-Term Energy Supply and Demand Outlook, adopted by the government in July 2015, envisioned a target share of 20–22 percent for nuclear power by 2030 (METI 2015), a target that was reconfirmed in the 2018 SEP (METI 2018) as well as the 2021 SEP (METI 2021). Japan's post-3.11 energy strategy stressed the continued importance of nuclear power, albeit at a lower level, and vowed to put safety first. Political promotion of nuclear power was a core feature of Japan's pre-3.11 nuclear politics, one seemingly reinstated by the Abe government in 2014.

However, it soon became apparent that political support after 3.11 was not quite the same as before. Japan's Federation of Electric Power Companies, called Denjiren, had severely opposed the nuclear phaseout option tabled in 2012 (Denjiren 2012) and highly welcomed the "promotion" of nuclear power in the 2014 SEP, which it regarded to be of "great significance" (Denjiren 2014a, 3). Pushing for the construction of more nuclear power plants traditionally was such an element of political support. Over the years, both Denjiren and Keidanren called for the construction of new nuclear reactors and for repowering, that is, replacing older smaller reactors with newer and larger ones (Economic News 2017; Denjiren 2013, 2; 2017, 2; 2018a, 3; 2019, 3; Keidanren 2019, 11). Abe also advocated for the construction of new nuclear power plants ahead of the election in 2012. However, asked about this after his election as prime minister, Abe neither confirmed nor denied such intentions, but rather ambiguously stated that this was not a matter to be decided immediately (Abe Shinzo 2013). Furthermore, the 2014 and 2018 SEPs contained only a backdoor provision to add reactors in order to reduce Japan's greenhouse gas emissions rather than a clear commitment. Hence, the Abe government was actually less supportive of nuclear power than previous LDP governments.

Much to the dismay of the nuclear industry, the Abe government provided less support despite otherwise influential industry associations' repeated insistence on it. Keidanren voiced its dissatisfaction over lackluster support by the Abe government by calling on them to "state their thorough commitment to the necessity of nuclear power" (*genshiryoku no hitsuyōsei nitsuite, teineina setsumei wo tsudsuketeikubeki*) (Economic News 2017). The electric utilities on their part voiced their dissatisfaction over the situation behind closed doors rather than in public. Their dismay also stemmed from their perception that the government and METI made TEPCO, and by extension all electric utilities, the scapegoat for the nuclear accident. Instead of apologizing for their involvement in the insufficient safety governance that enabled the accident, the LDP and METI attacked TEPCO, which eventually took the blame and apologized (interview with Japanese Energy Industry Expert, 2015). Both utilities and Keidanren were unsatisfied with government support for nuclear power, but only Keidanren voiced it openly.

The Abe government's shift in stance on nuclear power was connected to its desire to win the election in 2014 amid continued skepticism in the general public over nuclear power. To begin with, Abe avoided taking clear stances during the 2014 electoral campaign (Maeda 2015). Furthermore, the LDP adopted a position on nuclear power that was rather close to that expressed by voters in public opinion polls. In addition to the ambiguous nuclear policy position outlined above, the government supported a full market liberalization proposal, which electric utilities had—successfully—opposed in the 1990s. The rationale for utilities was that nuclear power promotion and a liberalized market were at odds with one another, because the latter required utilities to think like strategic actors in a competitive environment, which would discourage investments in nuclear power (Kikkawa 2012). Hence, the proposal for full market liberalization put additional pressure on electric utilities at a time when they were already dissatisfied with becoming the government's sole scapegoat for the nuclear accident and its lackluster support for nuclear power. As Hughes (2015, 204) aptly summed up the change in position, the Abe government "jettisoned its long-standing position of actively promoting the growth in nuclear power" in order to avoid nuclear power emerging as a liability during the 2014 election. Hence, a gap emerged between the political and administrative cornerstones of the nuclear iron triangle on the one hand and the nuclear industry, particularly utilities, on the other.

This came at a time when strong political and administrative support for nuclear power was essential as, without plans for starting up additional NPPs, tensions existed between the target share of 20–22 percent and the newly introduced safety regulation. Documents published by ANRE illustrated that the goal of 20–22 percent nuclear power hinges on nuclear power plants remaining in operation for more than forty years (ANRE 2015, 5). The idea to put a forty-year limit on the life span of nuclear power plants was something the LDP had opposed during the reform process in 2012. The NRA later introduced such regulation on its own. Hence, the 2030 target share for nuclear power was a government attempt to bring the NRA in line with its idea of safety regulation, that is, letting reactors run for sixty years, and a way to meet electric utilities' demands to restart nuclear power plants beyond the forty-year limit (Denjiren 2013, 2). Even though the 2014 and 2018 SEPs developed under the auspices of METI pressured the NRA to water down the newly introduced safety standards, the NRA has not relaxed the forty-year rule.

In a surprising turn, the government has backtracked on pressuring the NRA on this matter. In 2016, then METI minister Hayashi Motō stated that the 2030 nuclear power targets were rooted in the "expectation" (*mikomi*) that Japan's NPPs would have their life span extended beyond forty years in operation. Otherwise the share of nuclear power in the energy mix would be below 20 percent by 2030. Right afterwards, Abe clarified that the government would not restart reactors against NRA safety concerns just because of the 22 percent goal (Tokyo Shimbun 2016, 3). Thus, the Abe government admitted to the NRA standing in the way of achieving the government's 2030 nuclear energy targets, and it vowed to accept the safety agency's verdict. Such language also found its way into the 2018 and 2021 SEPs, which stress safety as a "top priority" and vow that

> judgment as to whether nuclear power plants meet the new regulatory requirements will be left to the Nuclear Regulation Authority (NRA) and in case that the NRA confirms the conformity of nuclear power plants with the new regulatory requirements, which are of the most stringent level in the world, GOJ will follow NRA's judgment and will proceed with the restart of the nuclear power plants. (METI 2018, 23; 2021, 8).

This language seeks to depoliticize nuclear power by making it a technocratic issue the NRA is responsible for, a strategy the Abe government adopted ahead of the 2014 snap elections (Hughes 2015, 203).

The seeming contradiction between the government's acknowledgement of potentially missing its nuclear policy goals and later reconfirming them reveals the balancing act the Abe government was trying to achieve. The fact that it took two years of negotiations within the Abe administration to agree on targets, shows how contested the future share of nuclear power was. The adopted numbers can be seen as symbolical in nature. An unnamed source in the Cabinet Office told a journalist that the goal of 20–22 percent nuclear power for 2030 served the sole purpose of showing continued government support for nuclear power and was by no means fixed (Asahi Shimbun 2015). Electricity targets were also an expression of uncertainty. Kikkawa (2012, 14–15), a member of energy policy deliberation councils, argued that the nuclear power share resulted from uncertainties related to renewable energy development, energy saving, and clean coal rather than being an independent variable in its own right. The share for nuclear power was determined by subtracting the share of other energy technologies from the demand assumed for 2030. Therefore, the government retained the 2030 nuclear power goal to show support for the nuclear industry, communicate political goals vis-à-vis the NRA, and serve as a placeholder in times of uncertainty about Japan's future electricity mix while at the same time seeking to depoliticize nuclear power by referring nuclear restart decisions to an independent regulator.

Over the years, a nuclear safety conversation unfolded between the first NRA chairman, Tanaka Shun'ichi, and the Abe government. It began with Prime Minister Abe's first New Year's Address in 2013, when he voiced the expectation that all safety screenings would be finished within three years (Abe Shinzo 2013). A few days later, Tanaka explained during a press conference that the NRA was doing everything in its power to undertake the safety review process in a swift manner, but that it was technically impossible to complete the task within three years (NRA 2013c). By 2019, the NRA had conducted fifteen safety screenings, far less than Abe had called for.

As the first safety screening of a nuclear power plant was nearing completion, the Abe government used the expression "certifying safety" (*anzensei wo tanposuru*). In response to a journalist asking whether the expression used by the Abe government meant that the NRA ensured risk-free nuclear power, Chairman Tanaka stated:

> We are conducting different compliance investigations from the point of view of lowering the risk as much as possible. . . . Since it creates misunderstandings when we speak of certifying safety, we use the term to conduct compliance investigations. . . . Even if you say it sounds different from what politicians are saying, and yes we are certainly using different terminology, I am actually not so sure there is such a big difference in reality. (NRA 2014b, 20)

Even though Chairman Tanaka avoided contradicting Abe directly by passing it off as a matter of word choice, he nevertheless refuted the idea of "certifying safety," thereby renouncing the notion of absolute safety.

On the occasion of the fifth anniversary of 3.11 in March 2015, Abe spoke about the importance of nuclear power for Japan. In his speech, he omitted further calls for swift reactor restarts, even though the NRA was far from completing all safety reviews, and stressed the NRA's "high degree of independence" (*takai dokuritusei*). Interestingly, he used NRA terminology about "compliance with regulatory requirements" (*kiseikijun ni tekigou*) when speaking about the NRA's safety standards (Cabinet Secretariat 2016), which was also used in the 2018 and 2021 SEPs. With time, the Abe government seemingly accepted the NRA as an independent regulatory agency and partly adopted its terminology.

Another potential avenue for political influence is through pronuclear parliamentarians forging close ties with the new safety agency in order to influence it.

NRA-Diet relations were partly regulated by the NRA Law, which obliges the NRA to submit its annual report to the Diet. On the occasion of Chairman Tanaka appearing before the Diet to provide expertise on nuclear power, a group of LDP politicians, led by former LDP nuclear safety reform champion and advocate of agency independence Shiozaki Yasuhisa, approached Tanaka. They handed him a proposal in which they criticized a lack of contact between the NRA and politicians, accused the NRA of isolating itself rather than acting independently, and asked the NRA to meet with politicians and other "people involved in nuclear power policy" (*kankeisha*) to listen to their concerns. In response, Tanaka politely refused the plea by citing time concerns and indicated that increasing contacts with politicians would take away time from pressing tasks (Amano 2015, 21). During the nuclear safety administration reform debate in 2011 and 2012, Shiozaki Yasuhisa had been a stern advocate of making the NRA as independent as possible and acted as a reform leader inside the LDP. The "isolation" criticism revealed that the NRA's idea of independence was not what the LDP had in mind when it pushed for an independent safety agency.

Along with parliamentarians, the NRA also kept a distance from government ministries, most notably METI and MEXT. For example, METI, and its Agency for Natural Resources and Energy, can request NRA members to provide technical expertise in their policy research councils. In such cases, NRA staff join in a consultative capacity. They give advice but refrain from getting involved in policymaking itself (interview with NRA Official, 2015). Such interactions were similar to Tanaka appearing before the Diet to answer technical questions without getting involved in policy discussions.

One might expect regular exchange with MEXT, as the NRA and MEXT share jurisdiction over two incorporated administrative agencies, namely the Japan Atomic Energy Agency (JAEA) and the National Institute of Radiological Sciences (NIRS). A contentious issue arose when the NRA took issue with the JAEA's safety management at the fast breeder reactor Monju, a facility at the heart of METI's nuclear fuel cycle ambitions. In response to repeated irregularities in how JAEA maintained the safety of Monju, the NRA announced its "recommendation" (*kankoku*) that "JAEA does not have the capacity to operate Monju safely." Should MEXT fail to find a new operator for Monju, it would lead to the facility being decommissioned by the NRA (NRA Commission 2015d, 4). With the government and MEXT unable to find a suitable operator, Monju was permanently shut down, dealing a huge blow to Japan's nuclear fuel cycle ambitions. In the decision-making process, the NRA held five meetings with JAEA representatives. The published meeting agendas, the reference materials, and lists of participants, which included no representatives from MEXT or METI. Apparently, the NRA resolved to publicly threaten to shut down Monju without involving either MEXT as the other ministry overseeing Monju or METI as the ministry promoting a nuclear fuel cycle with Monju at the heart of it.

In sum, the NRA stood its ground in the face of pressure regarding the nature and strictness of Japan's safety governance. As the NRA's representative, Chairman Tanaka seemingly always found a technical reason to justify the NRA's refusal of requests as a way to depoliticize his statements and to deflect criticism. Abe's request for restarts within three years was technically impossible, the call for increasing exchanges with proponents of nuclear power was incompatible with the NRA's already full schedule and could have caused further unwanted delays, and the idea of certifying safety was at odds with a scientific understanding of safety that acknowledges absolute safety as a myth. An official who had worked alongside him attested to Tanaka's impressive political smartness as he quickly adjusted to the shift from scientist to government executive and learned how to navigate the necessary procedures and protocol (interview with Preparation Office Member, 2020). Bringing up government pressure to restart nuclear power plants, a former NRA board member denied that the board felt any such pressure. The reason was that the fate of nuclear power plants, including whether to restart them, was METI's responsibility and the NRA was only in charge of assessing whether or not they met the new safety standards (interview with Former Board Member, 2019). With this division of tasks in mind and by quickly adapting to the political environment the NRA was working in, Chairman Tanaka managed to uphold the NRA's independence and nuclear safety standards while staying within protocol and avoiding open conflict.

Board Members' Appointments

A means of long-term influence and an opportunity to sway NRA verdicts is the appointment of NRA board members. Considering that the NRA Commission has the last say in all regulatory decisions, shaping the NRA board is possibly the most effective way of influencing the NRA. The appointment process, as established by the NRA Act (Article 7), requires appointment by the prime minister based upon obtaining Diet approval. Adding to the NRA Law, the NRA introduced staggered terms of offices for the first board members to ensure continuity in the board. It resolved that two board members' terms will end after two years, another two after three years, and the chairman's after five years.

Some appointments were controversial. In fact, the LDP and Kōmeitō had expressed their opposition to the first board when it was presented to the Diet in the fall of 2012. Unable to gain consent, the Noda government at the time appointed the five members on a preliminary basis without Diet approval. This gave the recently elected Abe government an opportunity to replace all five in early

2013. The Abe government refrained from replacing the whole first board, even though it had initially opposed the appointment of the board members selected by the DPJ government. Replacing them after a few months, however, would have undermined the LDP's commitment to an independent nuclear safety agency.

It was time for Prime Minister Abe to appoint new commissioners in 2014, 2015, and 2017. The terms of Shimazaki Kunihiko and Ōshima Kenzō ended in 2014. Commissioner Shimazaki had been a controversial commissioner due to his calls for fundamental reassessment of the maximum force an earthquake could strike a plant with, ultimately resulting in stricter safety standards. This delayed the restart of the Sendai nuclear power plant in Kyushu, which the operator Kyushu Electric had hoped to be swift. Shimazaki's role in the process prompted a Kyushu Electric executive to say that "Shimazaki made us suffer" (Asahi Shimbun 2014a). While activists and some NRA officials had hoped he would remain, an unnamed utility official called his replacement a "small victory" for the nuclear industry (Japan Times 2014a). Officially, Shimazaki refrained from throwing his hat in the ring again. If he had, it is questionable whether the Abe government would have reappointed him. Finding evidence for decisions not made is difficult, but in this case otherwise talkative interviewees' refusal to comment on the matter points in that direction. Replacing Shimazaki with the geologist Ishiwatari Akira[1] as commissioner for earthquake and tsunami countermeasures was a way for the Abe government to remove an unwantedly strict board member and to spare the industry more "suffering."

To replace Ōshima Kenzō, Prime Minister Abe appointed Tanaka Satoru[2] as the new commissioner for nuclear security and safeguards. Tanaka Satoru was controversial since many regarded him as too close to the nuclear industry. As a professor in Tokyo University's Department of Nuclear Engineering, Tanaka Satoru had long-standing ties with the nuclear industry and had actively promoted and supported nuclear power expansion in the past. He sat on an advisory committee of nuclear plant maker Mitsubishi FBR Systems Inc. (July 2007–June 2014); chaired a panel to evaluate vitrification technology at Japan Nuclear Fuel in Rokkasho, Aomori Prefecture (May 2009–March 2014); and received over ¥500,000 in fiscal year 2011 from a foundation linked to TEPCO (Asahi Shimbun 2014c). In the appointment process, the Abe government disregarded the transparency rules the DPJ had put in place in 2012 and did not disclose how much money Tanaka Satoru had received from the nuclear industry. Regardless of criticism and seven opposition parties opposing Tanaka Satoru's appointment, the Abe government pushed the decision through the Diet.

Once appointed, NRA neutrality guidelines for board members required Tanaka Satoru to disclose the amount of industry funding he had received in the previous years. He had received ¥6 million in research funds from the nuclear power plant operator J-Power and the plant manufacturers Hitachi and GE between 2004 and 2010. He had also received funding in the three years prior to nomination, for example from TEPCO Memorial Foundation, Hitachi GE, and Taiheiyo Consultant, an engineering firm. Critical voices warned that Tanaka Satoru joining the NRA board could undermine the independence of the nuclear safety regulator and make safety checks a formality (Japan Times 2014a; Asahi Shimbun 2014b).

The most important appointment decision came in 2017, when Chairman Tanaka Shun'ichi, age seventy-two, resigned at the end of his term. Some observers feared that Tanaka Satoru's appointment had been part of a government plan to make a figure close to the nuclear industry head of the NRA (interview with Japanese Nuclear Industry Expert, 2016). There was reason to believe that the Abe government in fact sought to name Tanaka Satoru the new chairman due to his age and education. With a PhD from the University of Tokyo, which stood at the top of the university ranking in Japan, he could have been a likely candidate. After Tanaka Shun'ichi, he was the oldest board member. Thus, his seniority and education made Tanaka Satoru a likely candidate for a chairman with close industry ties. Tanaka Satoru succeeding Tanaka Shun'ichi would have been a clear sign of political influence by the Abe government.

Instead, Fuketa Toyoshi was appointed to succeed Chairman Tanaka. Fuketa, reappointed as the commissioner for nuclear safety in 2015, was the only other board member to remain of the first board appointed by the Noda government. As such, Fuketa clearly understood the aims Tanaka Shun'ichi had pursued as chairman. He had also been instrumental to enforcing thorough safety checks as commissioner for nuclear safety. Fuketa was named acting vice chairman in 2014, a clear indication he was being groomed as Tanaka's successor. Fuketa and Tanaka shared a history of working together at the JAEA before joining the NRA. While the JAEA has been criticized for its involvement in the nuclear village, both Tanaka and Fuketa showed a strong willingness to foster safety governance based on independence, transparency, and neutrality (discussed in chapter 3) and even declared their former employer JAEA unfit to operate Monju. Fuketa would remain chairman until 2022. For the time being, Tanaka also remained tied to the NRA as an adviser. Naming Fuketa as the new chairman and retaining the first chairman as an adviser ensured continuity in the NRA board and the safety agency's governance approach.

The appointment process is essentially a black box. Officially, the prime minister chooses a candidate. In practice, senior NRA officials make a list, which they pass to the MOE, which in turn shares it with the Prime Minister's Office, where a sole candidate is selected to then be presented to the Diet for approval (interview with NRA Official, 2015). Fuketa must have been on the NRA's list of potential successors put together by the NRA and passed to the MOE.

A few factors point to senior NRA and MOE officials pushing for the appointment of Fuketa. To begin with, Fuketa enjoyed the support of the incumbent chairman Tanaka. As a senior NRA official recalled, those working in the NRA were not surprised by the decision in favor of Fuketa, sensing that Tanaka wanted Fuketa to become his successor (interview with NRA Official, 2018). Upon inquiry as to whether he had pushed for Fuketa to become his successor, a former board member explained that he had hardly done a thing and that there was no need. At the time, everyone understood that Fuketa was the best choice due to his extensive expertise. While any change in chairman marks a potential crossroads, Fuketa was also the best choice to ensure continuity, a view that was widely supported, including by the government (interview with Former Board Member, 2019).

In addition to Chairman Tanaka, Fuketa most likely had strong backing from the MOE. The MOE's administrative vice minister, the highest nonpolitical position within the ministry, at the time was Morimoto Hideka. Morimoto had been a member of the Cabinet Office reform team that designed the NRA. Together with other reform team members, including board members, he moved into the NRA, where he held a senior position until 2014. Therefore, he had been working closely with Tanaka and Fuketa to bring their shared vision of independent and transparent safety governance to life, making him another likely advocate for Fuketa as the next chairman. In this case, the common practice of promotion based on seniority gave way to an apparent determination to choose a trusted successor to continue the work that former reform team members had started together.

The 2017 change in chairman was a litmus test for political influence on the NRA. If the appointment of Tanaka Satoru indeed had been an attempt to make someone with closer industry ties chairman, it failed. Instead of a political appointee to bring the agency more in line with government views, someone from within the board was elected to continue the governance path the NRA had set out on. Similarly, Tanaka Satoru's previously close industry ties failed to translate into influence over safety checks or the NRA's independence. The reasons for this included the NRA board's unanimous decision-making procedures, limiting each board member's influence, and nuclear safety commissioner Fuketa's strict stance on nuclear safety. Thus, the NRA passed the litmus test of withstanding political influence via the appointment process.

Overall, interactions between the Abe administration and the NRA illustrate the NRA's ability to defend its independence as a regulatory agency. Over time, the Abe government's early stance of pushing for swift restarts, announcing the continued promotion of nuclear power, and pressuring the NRA to relax safety standards gave way to increasing constraint. As the NRA defeated attempts to bring it in line with pre-3.11 governance practices, the government eased the pressure, acknowledged the NRA as a hurdle in achieving policy targets, and held the NRA up as an example of independent and thorough safety governance. Similarly, the government appointed the NRA's preference for the next chairman.

The Abe government attempted to walk a fine line of supporting nuclear reactor restarts while avoiding the impression of undermining the NRA's independence. There were expectations that the Abe government would pressure the NRA as its nuclear safety governance came to stand in the way of achieving the government's nuclear policy goals. However, the overarching reform goal to regain domestic and international trust, repeated like a mantra during reforms in 2011–12, curtailed the government in its attempts to pressure the NRA. In order to regain trust lost, the LDP sought to make a credible commitment to independent and stringent nuclear safety governance. Making such a credible commitment did not allow for overt pressure on the NRA, especially as the 2016 IAEA Safety Standards for People and the Environment, revised in response to the Fukushima nuclear accident, came to include "freedom from political pressure" (IAEA 2016a). While some LDP politicians complained about the NRA's lack of attention to political concerns, evaluations by the IAEA lent legitimacy to this approach by lauding the NRA for its strong independence, including keeping its distance from politicians.

Furthermore, the Abe government sought to depoliticize nuclear power ahead of the 2014 election by making it a technocratic issue. It did so by relaying responsibility for whether a nuclear reactor was safe enough for a restart to the NRA. Examples of this include omnipresent language that stresses that restarts will only take place once the independent NRA has decided about a plant's safety. This was part of "a series of strategic choices made by the Abe government prior to election that may have helped insulate it from the potential for energy policy to be an election-defining public policy issue" (Hughes 2015, 203). Hence, the government retained the 2030 nuclear power goal of 20–22 percent to show some support for the nuclear industry and to communicate political goals vis-à-vis the NRA, but it was limited in its ability to overtly pressure the NRA as it was seeking to regain trust lost and to depoliticize nuclear power by referring nuclear restart decisions to an independent regulator.

Dethroned Electric Utilities

By refusing to yield to political pressure, the NRA created a radically different environment for electric utilities operating nuclear power plants. Utilities used to operate under "reciprocal consent" (Samuels 1987), meaning that the Ministry of

Economy, Trade, and Industry (METI, formerly MITI) and politicians were seeking utilities' consent to proposed measures before implementing them. Furthermore, METI and safety regulators refrained from challenging electric utilities, which gave them the power to largely decide on how much to invest into accident countermeasures and into updating existing NPPs in line with new safety standards, a process called backfitting. The NRA, however, enforced broadly applicable and stringent safety standards, including mandatory backfitting for nuclear reactors and accident countermeasures (discussed in chapter 4). Notwithstanding room for improvement in safety standards, especially risk assessments, top-down safety regulation and forcing utilities to backfit put an end to "reciprocal consent," at least with regard to nuclear safety standards and safety investments.

Once new safety standards went into effect in July 2013, electric utilities were invited to submit requests for safety reviews of nuclear reactors. Within one month, utilities submitted a total of twelve requests. Shortly after, Denjiren announced the intention of electric utilities to comply with the NRA safety review, to gain local approval for restarts, and to restart nuclear reactors one by one (Denjiren 2014b, 3). Joining the Abe government in pushing for swift restarts, Denjiren implored the NRA to conduct efficient investigations and to make judgments promptly (Denjiren 2014c, 2; 2014b, 3; 2014a, 2). Within roughly two years, utilities followed up their initial safety review requests with another fourteen and one more in 2018. Including two reactors under construction, by 2021, utilities had requested safety checks for a total of twenty-seven commercial nuclear reactors as well as a safety review of the Rokkasho reprocessing plant, a facility for Japan's nuclear fuel cycle. Judging from Denjiren pushing for swift safety checks and utilities swiftly putting nuclear reactors forward for safety checks, utilities were initially eager to bring nuclear reactors back into operation.

For the safety review, the NRA established a council titled Review Meetings on Conformity to the New Regulatory Requirements for Nuclear Power Plants. Falling under the NRA's transparency guidelines, review meetings were streamed live and a recording, meeting minutes, and documents used were publicly available on the NRA website. Consultations between the NRA and electric utilities, where the NRA explained safety issues in more detail, were not recorded on video, but the NRA published the date, time, participants, and topic under discussion. Utilities showed little enthusiasm for communication limited to public meetings and short consultations. Unsatisfied with their involvement in decision-making processes, they called on the NRA to intensify its communication with the utilities (Denjiren 2016, 3). Holding all discussions in front of the public eye was not easy at first for the NRA board members and officials, but the burden was probably largest for the utilities. Since transparency mattered a great deal to the NRA, however, it was a question of utilities either getting used to it or being unable to receive the

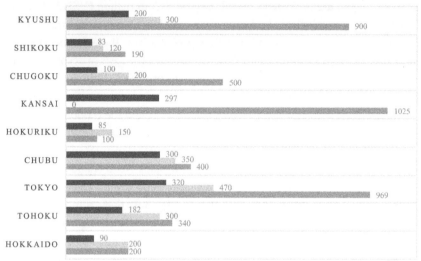

FIGURE 5.2. Costs of refurbishing by electric utility (in billion yen), created by the author based on Tokyo Shimbun (2015a) and Asahi Shimbun (2019a) data (Koppenborg 2021)

safety screenings necessary for nuclear reactor restarts (interview with Former Board Member, 2019).

New safety standards turned out to significantly raise technical safety costs. Asked by the *Asahi Shimbun* about the investments they were expecting to make to refurbish and restart reactors undergoing safety checks, utilities' estimates have gone from ¥1.7 trillion (roughly US$16 billion) in January 2013 to ¥5 trillion (a little less than US$50 billion) in July 2019 (Asahi Shimbun 2019a). Figure 5.2 shows a breakdown of refurbishing costs for each utility. Note that KEPCO was unable to provide an estimate in 2015. Tōhoku Electric was able to give an estimate only for the Onagawa plant in 2019, while previous numbers included anticipated costs for Onagawa and Higashidori. Utilities cannot be sure of the costs unless refurbishments are finalized and approved. Throughout the safety review process, the NRA checks whether alterations to the NPP are up to standards and, if unsatisfied, orders more construction measures. Utilities' estimates illustrate that required safety investments have skyrocketed far beyond their initial expectations.

Not only do utilities need to invest into additional safety measures to restart nuclear reactors, at the same time they also have to shoulder the costs of maintaining both restarted and idled nuclear reactors as well as costs for decommissioning those deemed too expensive to restart. The overall costs to implement government-mandated safety measures, maintain facilities, and decommission all existing fifty-seven commercially operated nuclear power plants at nineteen plants in Japan are projected to reach around ¥13.46 trillion (US$123 billion). The amount was calculated by *Kyodo News* based on financial documents from electric utilities as well as interviews with representatives. These numbers include refurbishments, repairs, and labor costs but exclude the costs of decommissioning the damaged reactors Nos. 1 to 4 at the Fukushima Daiichi nuclear power plant (Kyodo News 2020).

Rising costs impacted utilities' cost calculations and the economic viability of operating nuclear power plants. For one, the number of review applications was still twenty-seven in September 2022, meaning that utilities submitted only one restart permit after 2015, that is for the Shimane No. 3 reactor under construction. Submitted applications reveal a clear trend toward younger reactors with a large generation capacity—in other words, those that exhibit the highest chance of generating enough profit in the future to make the necessary investments worthwhile. On the flipside of this approach were idled older reactors at the same site for which no safety screenings were requested. This is the case with Onagawa, Kashiwazaki-Kariwa, Hamaoka, Shika, Mihama, Ōi, Shimane, Ikata, and Genkai. Exceptions were Sendai and Takahama, where all reactors on site underwent safety checks. A look at the reactors for which utilities requested safety reviews revealed that profitability was a major consideration in the decision to attempt a restart.

In 2015, utilities began shutting down nuclear reactors. The six reactors at the crippled Fukushima Daiichi plant were the first to go into permanent shutdown, followed by Tsuruga No. 1 after the NRA found it to sit atop an active earthquake fault line and refused an operating license for a restart. Between 2015 and 2019, electric utilities announced the permanent shutdown of fourteen nuclear reactors, eight of which were older reactors with an output below 600MW. Older reactors require more refurbishment to meet safety standards and they have less electricity output and remaining runtime to generate the revenue to make such investments worthwhile. After 3.11, the number of available commercial nuclear power plants decreased significantly from fifty-four in 2011 to thirty-three by 2022, plus two reactors under construction.

The decommissioning of nuclear power plants poses a challenge to Japan's nuclear energy policy. Japan's electricity targets for 2030 assume two things: a stable electricity demand, which can only be achieved with significant energy efficiency improvements and energy savings efforts (METI 2015), and the restart and lifetime extension of most nuclear power plants (ANRE 2015). In other words, government targets require utilities to refurbish and restart all remaining nuclear reactors and navigate the demanding procedures required to gain a twenty-year extension to the operating license. Investment decisions of electric utilities will largely determine whether the Japanese government can reach its 2030 target of generating 20–22 percent of electricity used in the country from nuclear power.

Two scenarios illustrate the impact of utilities' investments decisions on Japan's future electricity mix. Both assume a stable electricity demand, as the government's 2030 electricity targets do. The first scenario describes utilities only refurbishing and restarting the twenty-seven reactors put forward for safety checks by 2022, including two under construction. In the second scenario, utilities will seek restarts of thirty-five reactors, including all currently available thirty-three and two reactors under construction. Table 2 below compares the two scenarios, beginning with the installed electricity generation capacity as a percentage of that available in early 2011, which covered about 29 percent of Japan's electricity demand. The table shows the technically possible and realistic nuclear power share by 2030, taking into account that an operation rate of 100 percent is not the norm for nuclear power plants. They usually operate at somewhere between 80 and 90 percent of their rated capacity, resulting in standard operating procedures at about 85 percent capacity. The twenty-seven reactors under safety review can provide about 14 percent of Japan's electricity mix by 2030, far from enough to reach the government's goal of 20–22 percent nuclear power. Even if electric utilities restarted all thirty-five reactors, it would barely be enough to meet the 2030 targets. Unless electric utilities, and the government for that matter, commit to building new NPPs, Japan's share of nuclear power in 2030 will most likely remain below the target share.

Utilities' reluctance to seek more restarts stood in stark contrast to their earlier push for a return to nuclear power. With time, rising technical refurbishment costs and lengthy safety reviews presented a financial challenge for electric utilities. One month after Denjiren declared that "it is essential to build a sustainable earnings structure and to put the business back on track" (Denjiren 2015a, 1), it announced different utilities' decision to permanently shut down a total of five older and smaller reactors (Denjiren 2015b, 2). This marked a turning point. Thereafter, Denjiren refrained from pushing for swift restarts and pledged to "assure an even higher dimension of safety that exceeds the regulatory framework"

TABLE 2 Impact of utilities' investments decisions on Japan's nuclear power targets for 2030

NUMBER OF REACTORS	INSTALLED CAPACITY (% OF 2011 CAPACITY)	SHARE IN 2030 ELECTRICITY MIX (%)	
		100% OPERATION RATE	85% OPERATION RATE
27	60	16	14
35	73	21	18

(Denjiren 2018b, 1; 2019, 3). Parallel to decommissioning another five older and smaller reactors, it shifted to stress "safety and economic efficiency" (Denjiren 2018a, 3; 2019, 3). As technical safety costs rose and utilities faced tough business decisions, an apparent shift in strategy took place where utilities discarded the idea of restarting all reactors in line with government plans in favor of a more business-oriented strategy with strategic investments.

Concomitant with the shift towards strategic investment, changes in the landscape of nuclear power plant operators took place. Out of the twenty-seven requests for safety checks in order to bring nuclear reactors back in operation, eighteen came from utilities operating only small numbers of nuclear reactors. For example, Hokkaido Electric attempted to restart all of its three reactors and Kyushu Electric four out of five. In contrast, TEPCO, which operated seventeen reactors until 2011, one-third of all nuclear reactors in Japan, only applied for two safety checks. KEPCO, the second-largest electric utility with eleven reactors in operation prior to 3.11, applied for restarts of seven while decommissioning four. KEPCO has traditionally generated a large share of its electricity using nuclear power, but more recently it has been trying to remove nuclear power from its portfolio (interview with Japanese Energy Industry Expert, 2015). These changes signaled a shift away from the traditional nuclear powerhouses TEPCO and KEPCO, which once operated half of Japan's nuclear power capacity between the two of them, to mainly smaller utilities operating a small number of reactors each.

While KEPCO appears to be working on an exit strategy, TEPCO may be leaving the nuclear power business by default. To begin with, the decision to retire all six reactors at the Fukushima Daiichi NPP, instead of only the four damaged ones, at once removed one-third of the reactors in TEPCO's portfolio. In its decision to also permanently shut down the four reactors at Fukushima Daiichi, TEPCO took into account that the governor of Fukushima Prefecture repeatedly stressed his opposition to restarting any nuclear power plants in Fukushima Prefecture (TEPCO 2019a). TEPCO has struggled to gain local consent to restarting reactors since 3.11. Even though not legally required, it is customary to seek local government approval. Out of TEPCO's remaining seven reactors at Kashiwazaki-Kariwa in Niigata Prefecture, the two youngest and largest were undergoing safety reviews. Successive governors of Niigata Prefecture have opposed TEPCO's restart attempts before the Niigata prefectural investigation into the March 2011 accident is completed, which can serve as the basis for a decision about TEPCO's suitability as an operator. Furthermore, the major of Kashiwazaki town, Masahiro Sakurai, requested TEPCO to permanently shut down the five reactors not undergoing safety reviews. In response, TEPCO offered to scrap at least one of the older reactors within five years of restarting no. 6 and no. 7 (TEPCO 2019b), a timeline Sakurai accepted. Depending on local government decisions, TEPCO may have to scrap at least one more rector at Kashiwazaki-Kariwa—or give up entirely its last remaining commercial nuclear power plant. Construction work on its only new reactor has been on hold since March 2011 (TEPCO 2020). Hence, TEPCO, the first electric utility in Japan to pursue the nuclear power business and a once proud powerhouse of the nuclear industry, has thus far failed to bring a single reactor back into operation.

Changes in the landscape of nuclear power plant operators and their investment strategies took place in the context of an increasingly difficult business environment for electric utilities. After 3.11, they were faced with lackluster political support for nuclear power, a liberalized electricity market, exploding safety costs, and a safety agency unresponsive to utilities' (and politicians') concerns. This stands in stark contrast to the pre-3.11 business environment, where nuclear industry actors enjoyed a powerful position within the nuclear triangle that allowed them to influence nuclear policymaking and safety regulations and where power plants generated a steady stream of revenue for the plant operators.

Galvanized Antinuclear Protests

Utilities that were willing to make necessary investments and were able to gain local consent still faced an acceptance hurdle. When large antinuclear demonstrations, sparked by the March 2011 nuclear accident, subsided after 2012, it was not a sign of recovering public approval of nuclear power. Rather, public opinion remained skeptical with a majority still opposed to nuclear restarts in 2019, as shown in figure 5.1 above.

The NRA's transparent safety governance approach keeps public opinion informed and invites domestic and international scrutiny of nuclear safety. Scrutiny of NRA activities came from, among others, established antinuclear NGOs in Japan. Japan's oldest antinuclear nongovernmental organization, the Citizen Nuclear Information Center (CNIC), established in 1975, regularly posted critical coverage of NRA safety screenings under a designated header on its website.[3] Similarly, Greenpeace Japan regularly posted press releases in both English and Japanese on nuclear safety issues. It criticized the NRA for safety inspections using the old checklist system (Greenpeace Japan 2016b, 2016c), insufficient safety assessments before extension of reactors' lifetime beyond forty years (Greenpeace Japan 2016a; CNIC 2015), and risk assessments that excluded unverifiable risks, such as an earthquake fault line with unknown activity status (interview with Ban Hideyuki, 2015; Vande Putte, Ulrich, and Burnie 2015). Both CNIC and Greenpeace regularly drew attention to areas where the NRA should have been more stringent in their view.

For citizens, one route to oppose nuclear power plants has always been to appeal to the courts. Between 1973 and 2010, there were a total of 15[4] class-action lawsuits filed against nuclear power plants. Citizens brought lawsuits in district courts within the same prefecture and the outcome was usually the same. In the case of a lawsuit against the Shika No. 2 reactor, where the Kanazawa District Court ruled in favor of the plaintiffs, the Nagoya High Court later overturned the decision. All but one district court dismissed lawsuits in the first instance and higher courts have upheld the decision (CNIC, pers. comm.).

In 2011 and 2012, new networks of antinuclear civil society actors emerged (Wiemann 2018, Brown 2018). For example, Japanese lawyers who have been involved in class-action lawsuits for decades founded the Assembly of Japan's Lawyers for a Nuclear Phaseout in 2011 in order to improve knowledge exchange and improve coordination of activities (Bengodan 2020). After demonstrations subsided, antinuclear civil society activity returned to the previously prevalent local or NIMBY protests, albeit with new networks connecting them.

After 3.11, efforts to restart nuclear power plants stimulated a wave of cases brought before courts all over Japan. Between 2011 and 2020, there were thirty lawsuits against commercial nuclear power plants.[5] In the nine years following 3.11, citizens flooded operators and the regulators with twice as many antinuclear lawsuits than in the four decades prior. Table 3 gives an overview of post-3.11 lawsuits, organized by the targeted nuclear power plant, and provides information about the review status, courts involved, plaintiffs' demands, and respective rulings. To distinguish cases that target NRA decisions, lawsuits brought before safety reviews began in July 2013 are depicted in italic font. Please note that the data in this section are taken from the table unless otherwise specified.

Most attempts to restart nuclear power plants have been challenged by citizens' class-action lawsuits. Out of the thirty lawsuits between March 2011 and early 2020, fifteen were filed before the new safety standards went into effect and another fifteen were filed after the NRA began conducting safety checks. Of the twenty-seven nuclear reactors seeking restarts, twenty-three, or 85 percent, faced legal action. Exceptions, at least for the time being, were Shimane No. 2, Onagawa No. 2, Higashidōri No.1, and Tsuruga No. 2. Most of the fifteen reactors that passed safety checks, the exception being Kashiwazaki-Kariwa Nos. 6 and 7, faced lawsuits. Hence, citizens launched class-action lawsuits against 85 percent of reactors seeking restarts and against 86 percent of positively concluded safety checks.

By early 2020, six lawsuits had come to a conclusion with different outcomes. In all six cases, at some point there were court decisions in favor of citizen plaintiffs

TABLE 3 Lawsuits filed between 2011 and 2020, created by the author based on information about safety review applications and approval from the NRA Website and CNIC (CNIC 2020) data on lawsuits (Koppenborg 2021)

NUCLEAR POWER PLANT SAFETY REVIEW REQUESTS (NRA APPROVAL, IF APPLICABLE)	LAWSUIT	DISTRICT COURT	HIGHER COURT
Tomari No. 1–3 July 2013	*Tomari 1–3*	*Sapporo District Court, decommissioning on 2011.11.11*	
Ōi No. 3 & 4 July 2013 (May 2017)	Ōi 3–4	*Ōi District Court, halt operations on 2012.06.12*	
	Ōi 1–4	*Kyoto District Court, injunction on 2012.11.29*	
	Ōi 3–4	*Fukui District Court, injunction on 2012.11.30, injunction granted on 2014.05.21*	*Nagoya High Court Kanazawa Branch, defendants' appeal on 2014.05.22, plaintiffs' (partial) appeal on 2015.05.30 Lifted injunction and rejected plaintiff's demand on 2018.07.04*
	Ōi	Ōtsu District Court, injunction on 2013.12.24	
	Ōi 3–4	Fukui District Court, injunction on 2014.12.05, injunction granted on 2015.04.14 Fukui District Court, defendant's appeal on 2015.04.17, injunction lifted on 2015.12.24	Nagoya High Court Kanazawa Branch, plaintiffs' appeal on 2016.01.06, repudiated injunction in March 2016
Takahama No. 3 & 4 July 2013 (February 2015) Takahama No. 1 & 2	Takahama 3–4	Fukui District Court, injunction on 2014.12.05, injunction granted on 2015.04.14 Fukui District Court, defendant's appeal on 2015.04.17, injunction lifted on 2015.12.24	Nagoya High Court Kanazawa Branch, defendant's appeal on 2016.01.06, repudiated injunction in March 2016
March 2015 (April 2016)	Takahama 3–4	Ōtsu District Court, injunction on 2014.12.05, injunction granted on 2016.03.09 Defendant's appeal on 2016.03.12, appeal rejected on 2016.07.12	Osaka High Court, defendant's appeal on 2016.07.14, injunction lifted on 2017.03.28
	Takahama	Ōtsu District Court, injunction on 2013.12.24	
	Takahama 1–2	Nagoya District Court, halt lifespan extension on 2016.04.14	

(continued)

TABLE 3 *(continued)*

NUCLEAR POWER PLANT SAFETY REVIEW REQUESTS (NRA APPROVAL, IF APPLICABLE)	LAWSUIT	DISTRICT COURT	HIGHER COURT
	Ikata 1–3	*Matsuyama District Court, injunction 2011.12.08*	
	Ikata 2–3	Hiroshima District Court, injunction filed on 2016.03.11, injunction rejected on 2017.03.30	Hiroshima High Court, plaintiffs' appeal on 2017.04.13, injunction granted on 2017.12.13
Ikata No. 3 July 2013 (July 2015)	Ikata 2–3	Ōta District Court, injunction on 2016.09.28	
	Ikata 3	Yamaguchi District Court Iwakuni Branch, injunction on 2017.03.03, injunction rejected on 2019.03.15	Hiroshima High Court, plaintiffs' appeal on 2019.03.15, injunction granted on 2020.01.17
	Ikata 2–3	Yamaguchi District Court Iwakuni Branch, injunction on 2017.12.27	
Sendai No. 1 & 2 July 2013 (September 2014)	*Sendai 1-2*	*Kagoshima District Court, injunction and ban operations on 2012.05.30*	
	Sendai 1-2	Fukuoka District Court, revoke approval of installation changes on 2016.06.10	
Genkai No. 3 & 4 July 2013 (January 2017)	Genkai 1-4	Saga District Court, injunction on 2012.01.31	
	Genkai 1–4	Saga District Court, injunction on 2011.12.27	
	Genkai 2-4	Saga District Court, injunction on 2011.12.27	
	Genkai 1–4	Saga District Court, ban operations on 2013.11.13	
Kashiwazaki-Kariwa No. 6 & 7 September 2013 (December 2017)	Kashiwazaki-Kariwa 1-7	Niigata District Court, injunction on 2012.04.23	
Shimane No. 2 December 2013 Shimane No. 3 August 2018	None Shimane 3	*Matsue District Court, revoke construction license, injunction on 2013.04.24*	

Reactor (date)		Court action
Onagawa No. 2 December 2013	None	
Hamaoka No. 4 February 2014	Hamaoka 3–5	*Shizuoka District Court Hamamatsu Branch, permanent shut down and ban on operations on 2011.05.27*
Hamaoka No. 3 June 2015	Hamaoka 3–5	*Shizuoka District Court, halt operations on 2011.07.01*
Tōkai No. 2 May 2014 (September 2018)	Tōkai 2	*Mito District Court, revoke operating license, injunction on 2012.07.31*
Higashidōri No. 1 June 2014	None	
Shika No. 2 August 2014	Shika1-2	*Kanazawa District Court, injunction on 2012.06.12*
Ōma December 2014	Ōma construction license	Tokyo District Court, revoke construction license, cease-and-desist letter on construction on 2014.04.03
Mihama No. 3 March 2015 (October 2016)	Mihama Mihama 3	Ōtsu District Court, injunction on 2013.12.24 Nagoya District Court, revoke operation extension on 2016.12.09
Tsuruga No. 2 November 2015	None	

seeking to halt a nuclear reactor. In two lawsuits against Ōi Nos. 3 and 4 and another two against Takahama Nos. 3 and 4, decisions in favor of plaintiffs by the district courts in Fukui and Ōtsu were later overturned by the Nagoya High Court and the Osaka High Court. Higher courts acting as gatekeepers in case district courts were receptive to antinuclear opposition was in line with pre-3.11 lawsuits.

Unexpectedly, a high court overturned a district court decision with a ruling in favor of plaintiffs. In two cases, the Hiroshima High Court ordered a halt of Ikata No. 3, first in 2017 and then again in 2020. In doing so, it overruled the Hiroshima District Court and the Yamaguchi District Court decisions, respectively, that had allowed operators to keep the reactor in operation. Consequently, a mixed picture has emerged. While many higher courts apparently still ruled more in favor of utilities and the state, they were not automatic gatekeepers for government and utilities anymore.

The effectiveness of antinuclear lawsuits depends on the receptiveness of courts to the plaintiffs' cause. With regard to NRA safety standards, rulings differed widely between judges. In April 2015, the Fukui District Court ordered an injunction against Takahama Nos. 3 and 4, which had just cleared the NRA safety screening. The presiding judge, Higuchi Hideaki, criticized the data used for the earthquake risk assessment and found fault with NRA safety standards as "not strict enough and not appropriate" (*yuruyakade fugōri*) (Japan Times 2015b, 1; Tokyo Shimbun 2015c). In the same month, the Kagoshima District Court dismissed an injunction against the Sendai Nos. 1 and 2 reactors, which had also received a green light by the NRA. The presiding judge, Maeda Ikumasa, praised new safety standards as based on the "newest scientific knowledge" (*saishinde kagakuteki*) and "appropriate" (*gōriteki*) (Japan Times 2015a). Within the span of one month, two judges came to opposite conclusions about the adequacy of Japan's new safety requirements.

The six court decisions in favor of citizen plaintiffs trace back to three judges. Higuchi, the beforementioned chief judge at the Fukui District Court, granted three injunctions, twice against Ōi Nos. 3 and 4 and once against Takahama Nos. 3 and 4, which forced operations to stop. With Fukui Prefecture at the heart of nuclear power generation in Japan, a judge with a critical attitude towards nuclear power would have been a major hurdle for reactor restarts in Fukui. At first, the operator of both nuclear power plants, KEPCO, tried to have Judge Higuchi removed (Tokyo Shimbun 2015b) but failed to do so. Soon after the rulings against KEPCO, Higuchi was transferred to a different court and, following an appeal by KEPCO, the new chief judge lifted the injunctions. In an interview after his retirement, Higuchi rejected speculations about his transfer being a punishment for repeatedly ruling against an electric utility. At the same time he encouraged younger judges to make independent judgements rather than to rely on precedents (Asahi

Shimbun 2018), which equaled a call for more rulings in favor of citizens' challenging nuclear reactor restarts on safety grounds. The Hiroshima High Court Judge Kazutake Mori, who surprisingly overruled district courts' decisions and forced Ikata No. 3 to halt operations in January 2020, retired that same month. Thus, at the time of writing, two of three receptive chief judges had retired, and it remained to be seen how many would follow in their footsteps.

Injunctions (*karishobun*) have played an increasing role in the judicial struggle over nuclear safety and reactor restarts. Seeking an immediate halt of operations via an injunction emerged in the 1980s as a new addition to citizen plaintiffs' toolbox. About half of the pre-3.11 class-action lawsuits sought an injunction, and the other half targeted the operating license granted by the state. Some of the thirty lawsuits filed since 2011 sought a ban on operations or construction. The majority, twenty-one lawsuits in total, sought an injunction (CNIC, pers. comm.). An injunction can be a powerful tool. If granted by the court, an injunction idles a nuclear power plant until the decision is overruled by a higher court.

Injunctions can pose a significant hurdle to operators seeking to restart nuclear reactors, even if the injunction is eventually overruled. For example, the Takahama No. 3 and No. 4 reactors in Fukui Prefecture received a green light by the NRA in February 2015, but in April 2015 the Fukui District Court ordered an immediate halt to operations via an injunction. Once the same court lifted the injunction in December, KEPCO began the process of restarting the nuclear reactors. After only a few days in operation, however, Takahama Nos. 3 and 4 had to be shut down again following an injunction ordered by the Ōtsu District Court in March 2016. The decision was ultimately overruled by the Osaka High Court in March 2017. For the operator, KEPCO, this meant a significant delay in the restart process. At the time the two reactors received a green light, they had seven years until reaching the forty-year life span limit, which necessitates another safety check to renew the operating license. Due to injunctions, the two Takahama reactors had two of seven remaining operating years removed by injunctions that were later lifted.

Takahama was not the only nuclear power plant to face a legal challenge from outside its host prefecture. New safety requirements require local governments to draw up mandatory emergency response plans for a precautionary action zone (PAZ), a thirty-kilometer radius around a nuclear power plant. In many cases, this means plant operators need to work together with local authorities outside their host prefectures to prepare evacuation plans. The second injunction against Takahama Nos. 3 and 4 was imposed by the Ōtsu District Court in Shiga Prefecture, located next to Fukui prefecture, on the grounds that KEPCO failed to provide necessary information on safety- and evacuation-related issues (Asahi Shimbun 2016a). Similar developments were visible with Ōi[6] facing lawsuits in Fukui prefecture and neighboring Shiga prefecture as well as Ikata facing a class-action lawsuit in Ōita, a prefecture falling within the thirty-kilometer radius around the plant. As a result of the NRA's expanded emergency response regulation, more courts were drawn into legal battles over nuclear restarts.

Expanded emergency response regulation also drew a local government into the nuclear power struggle. In a first, the mayor of Hakodate city in Hokkaido brought suit against the completion of the Ōma nuclear power plant in Aomori Prefecture. Hakodate city is only twenty-three kilometers away from Ōma power plant, just across the Tsuruga Strait, and thus falls within the thirty-kilometer PAZ. In the ongoing lawsuit, Hakodate city criticized the plant operator, J-Power, for failing to provide necessary information for emergency arrangements. Furthermore, the Hakodate mayor challenged the notion that only the host prefecture needs to consent to a reactor restart instead of all prefectures affected in case of an accident (CNIC 2014). The mandatory evacuation planning zone of thirty kilometers even stimulated calls for a broader involvement of local authorities in the decision-making process leading up to a reactor taking up operations.

Most of the concluded lawsuits, five in total, challenged NRA safety reviews for a particular reason. In a lawsuit against Takahama Nos. 3 and 4, filed at the Fukui District Court, plaintiffs criticized risk assessments for excluding possible earthquake sources, resulting in insufficient earthquake safety measures (Japan Times 2015b; Tokyo Shimbun 2015c). In a similar lawsuit against Takahama Nos. 3 and 4 filed at the neighboring prefecture's Ōtsu District Court, plaintiffs challenged risk assessments on the grounds that they failed to take into account all risks facing the nuclear power plant (Asahi Shimbun 2016a). Likewise, citizens filed a class-action lawsuit against Ikata Nos. 2 and 3 over faulty earthquake ground motion figures used as the basis for the reactor's antiseismic design as well as underestimation of the risk posed by the Mount Aso volcano (Asahi Shimbun 2017; Japan Times 2020). These lawsuits highlight one nuclear safety area plaintiffs criticize as insufficient, namely risk assessments pertaining to earthquakes and volcanic eruptions.

This criticism of risk assessments can be seen as part of a larger struggle over dealing with the risks posed by Japan's many volcanoes and the earthquake fault lines that run underneath and near the archipelago on the Pacific "ring of fire." Discussed in more detail in chapter 4, the NRA follows a traditional risk assessment approach that focuses on risks that can be scientifically verified. Taking a strict stance on verified risks, it imposed considerable additional costs on electric utilities and shut down one NPP and a nuclear fuel cycle facility over safety concerns. As part of safety reviews, however, the NRA accepted utilities' risk assessments, which only included proven sources of risk and assumed that other potential but unverifiable hazard sources were negligible. Unverifiable risks included the likelihood of a volcanic eruption on Kyushu, which scientists were unable to predict as of yet, and the likelihood of uninspected earthquake fault lines off the coast of Japan causing an earthquake. By accepting such risk assessments, the NRA rejected a more precautionary approach of addressing all potential risk. Lawsuits challenging risk assessments thus protest the NRA's rejection of the precautionary principle and demand the safety agency address all potential risks.

After 3.11, new trends in class-action lawsuits and respective court decisions have become visible. Next to a significant increase in numbers, twice as many class-action lawsuits in less than a quarter of the time, there have been qualitative changes. One such change was the widespread use of injunctions, which began in the 1980s and took on a new dimension after 3.11 with the majority of lawsuits using injunctions to seek, at least temporarily, a shutdown. Another development was an expanded notion of "local" protests. Particularly the NRA's expanded emergency response regulation drew residents from neighboring prefectures into the political struggle over restarting nuclear power plants, resulting in the filing of lawsuits not only before courts in a plant's host prefecture. In doing so, NRA safety standards have expanded the metaphorical backyard of Not-In-My-Backyard (NIMBY) protests. Interestingly, they also led a city mayor to challenge customary "local" consent by demanding its expansion to include all prefectures affected in case of an accident. In addition, new safety requirements, such as a second license after forty years in operation, provided citizens with more opportunities to challenge a reactor's operating license. There were clear trends toward a higher number of lawsuits, more injunctions as a potentially powerful opposition tool, and an expanded notion of the terms "local" and "backyard." The level of courts' receptiveness to citizen opposition to nuclear power crucially depends on how receptive judges will be in the future to such legal challenges.

A Transparent and Independent Regulator as a Game Changer

The NRA presented a dilemma for the Abe government's ambitions to continue utilizing nuclear power. Initially, the Abe government attempted to pressure the NRA to abandon the forty-year life span, to conduct safety checks more quickly—thus less thoroughly—and to consult more with pronuclear actors, such as supportive parliamentarians and industry actors. Government pressure on the NRA to return to pre-3.11 business as usual had been identified as a challenge for the new safety agency's independence (Kingston 2014; Shadrina 2012). The NRA, however, fended off attempts to bring it in line with the nuclear village. In fact, the degree of independence and transparency the NRA attained was much higher than what the LDP had in mind when it supported the creation of an independent safety agency in 2012 in order to regain public trust in Japan's nuclear safety administration.

Theoretically, the NRA's independence could have been even more reason for the government to pressure the NRA into readopting previous safety governance practices. In practice, the LDP's previous mantralike repetition of calls for independence to regain domestic and international trust lost after 3.11, as well as its goal to depoliticize nuclear power, came to serve as a brake on the government's ability to pressure the NRA. In the context of a skeptical domestic and international audience, a credible commitment to nuclear safety forbade overt attempts to undermine the NRA's independence. Scrutiny by international actors, such as the IAEA, played an important role (discussed also in chapter 3), as Japanese public opinion would gain great reason to doubt the veracity of Japanese government statements about strong and independent safety regulation if the IAEA—or the US NRC for that matter—disagreed. As government pressure came to nothing, Abe shifted from a mixture of open pressure and praise to a strategy of mainly praising the NRA's independence and high safety standards coupled with indirect pressure via adopted policy goals. The Abe government showed increasing restraint vis-à-vis the NRA, which reflected the NRA's transparent and independent safety governance taking root.

Dealing with an independent nuclear safety agency put the government in a difficult position. On the one side, Keidanren, Japan's leading business association, and Denjiren, representing electric utilities, called for stronger government support, including the construction of new nuclear reactors. On the other side, the Abe government openly admitted to the NRA standing in the way of achieving governmental nuclear power goals for 2030. Despite this, the Abe government reconfirmed the same goals in 2018, and later the government of Prime Minister Yoshihide Suga in 2021, as a mixture of showing some political support to the nuclear industry, concealing uncertainty about Japan's future electricity mix, and communicating political goals to the NRA. The Abe government tried to balance different interests by stressing safety and regulatory independence while voicing support to the industry. However, it was not lending the same amount of political support to nuclear power almost a decade after 3.11 as it had done right after, let alone before, the accident. The LDP government abandoned calls for an increase in the share of nuclear power as well as the construction of new NPPs and imposed the burden of full retail electricity market liberalization on electric utilities. Unlike what terms such as Kingston's (2016) "Abenomics' fourth arrow" and Incerti and Lipscy's (2018) "Abenergynomics" suggest, the Abe government's political support for nuclear power became halfhearted.

For the electric utilities, top-down regulation by the NRA signaled a decline in power that was not received well. Before 3.11, electric utilities enjoyed plenty of state support as partners in both policymaking and implementation, alternatively dubbed "reciprocal consent" (Samuels 1987) or "private implementation of public policy" (*kokusakuminei*) (Kikkawa 2012). As part of this system, standards for older nuclear reactors in operation and costly accident countermeasures were largely left to the utilities to be implemented at their discretion, allowing them to determine how much to invest into nuclear safety. This practice contributed to the low operating costs of nuclear power regularly praised in Japan's Strategic Energy Plans, but it also led to the worst nuclear accident since Chernobyl, if not in history. Ensuing reforms achieved something electric utilities had successfully opposed prior to 3.11, safety requirements firmly in the hands of a state safety agency. The NRA turned around power relations between state regulator and industry.

As a result, Japan's nuclear safety governance shifted from reciprocal consent to a "regulatory state" (Levi-Faur 2011). This created a fundamentally different business environment for electric utilities. It went from "profitable and virtually risk-free" nuclear power expansion (Cohen, McCubbins, and Rosenbluth 1995, 193) in the heyday of Japan's nuclear power development to top-down safety regulation imposing massive technical safety costs. Utilities' operating profit margins, the proportion of revenue which remains after deducting costs, gradually decreased from 13 percent in 1986 to low single digits by 2011 (Scalise 2014, 96). Only as long as NPPs required no further safety investments, were they important assets to financially pressed utilities. The NRA, however, imposed massive additional safety costs amid increasing cost concerns following the retail market liberalization. This new business environment posed a financial challenge for utilities.

As electric utilities lost their position as partners in implementing safety measures, utilities ceased to make the investments necessary to achieve government energy policy targets. In 2013, when assumed necessary investments were comparatively low, Denjiren pushed for restarts. As safety investments required by the NRA continued to rise, electric utilities adapted to the costs imposed on it by the NRA by investing strategically. Utilities decommissioned older and smaller reactors and, instead, invested strategically into putting younger and larger reactors forward for safety inspection. Furthermore, local disapproval thwarted many of the utilities' intentions to restart reactors, especially for TEPCO. A study of restart attempts found certain criteria, such as age, size, or local consent, to be less important than regulators and courts in determining restart prospects (Aldrich and Fraser 2017, 454). The findings here show that these factors matter in utilities' decision whether to attempt a restart in the first place. By late 2019, barely enough commercial reactors remained to reach the 20–22 percent nuclear power share by 2030.

Next to binding safety regulation, increased transparency was a game changer after 3.11. The NRA's openness about the risks of nuclear power, and insistence on the fact that there is no such thing as absolute safety, thwarted any hopes to reestablish a new version of Japan's previous safety myth. At the same time, expanded safety regulation, and the conservative safety regulation stance that served the NRA well as a way to remain apparently neutral in Japan's nuclear power struggle, have brought more critics into the arena as evident from numerous class-action lawsuits challenging conservative risk assessments in particular. Meanwhile, continued domestic and international scrutiny worked in favor of the NRA by protecting it from pressure by the nuclear village. This is in line with earlier findings about high public interest and public mobilization increasing the chances of establishing an independent agency (Wilson 1989). Transparency served to protect the NRA's independence and helped retain a higher level of scrutiny than before.

Intense scrutiny persists and, with one exception, every NRA decision to grant an operating license has been challenged by citizens in court. Furthermore, NRA safety requirements provide more opportunities than ever to citizens seeking to challenge restart attempts. For one, the requirement for utilities to obtain a second license after forty years gives citizens a second opportunity to challenge nuclear reactor operation licenses. Second, the requirement for local governments within a thirty-kilometer radius to prepare evacuation plans gives courts in neighboring prefectures jurisdiction and citizens an opportunity to contest a restart attempt there. Many citizens filed class-action lawsuits before courts in adjacent prefectures. Consequently, antinuclear protests in Japan, usually described as local or NIMBY protests, have broadened in scope. Put differently, the NRA safety requirements galvanized the local protests already present before 3.11. As a result of the NRA expanding the symbolic backyard and providing more opportunities for citizens to challenge each nuclear reactor, nuclear power opposition evolved from NIMBY "site fights" (Aldrich 2010) into cross-prefectural reactor fights.

This new environment also exacerbated differences between plant manufacturers and operators. Already in the 2000s, the two nuclear industry groups saw relations sour. While manufacturers became exporters of nuclear power plants as part of Japan's "nuclear renaissance" with support from METI, utilities suffered from cost overruns and became less likely to strike a compromise with manufacturers, leading to the two industry groups losing their previously unified position (Hymans 2011, 178–79). Since then, the additional technical safety costs and social acceptance costs have affected the two industry groups unevenly, with utilities shouldering the brunt of the financial burden. Even though both industry groups' associations called for the government to support the construction of new NPPs, utilities refrained from purchasing new reactors from manufacturers and put ongoing construction projects, apart from two near completion, on hold. Disagreements became publicly visible in 2017, when Keidanren began calling on electric utilities to be more vocal in their support of nuclear power, but Denjiren refused to.

Since 2011, the contours of the nuclear village have become fuzzier. There have been instances of traditionally pronuclear group members speaking out against nuclear power. For example, Japan's former LDP prime minister Koizumi Junichirō turned into a vocal advocate of a nuclear phaseout. Another LDP politician, Kōno Tarō, criticized nuclear power and made YouTube videos questioning Japan's nuclear fuel cycle plans, but he ceased such activities after becoming a member of the Abe Cabinet. Then there is Shiozaki Yasuhisa, who was a champion of regulatory independence during the reforms in 2011–12 but later joined those pushing the NRA to pay more attention to political concerns, making it difficult to place him as an insider or outsider. Turning to Japan's three nuclear power plant manufacturers, Hitachi, Mitsubishi, and Toshiba, they moved into the heart of the global nuclear industry as part of the proclaimed "nuclear renaissance" in the 2000s. In 2018, however, Toshiba sold its stake in the US nuclear power business Westinghouse and had given up plans to build a nuclear power plant in the United Kingdom, indicating a move away from the nuclear power business. Among electric utilities, Japan's largest utilities serving the metropolitan areas Tokyo and Osaka plan to operate a drastically reduced number of reactors and have not tabled any new construction plans or continued ongoing construction projects. In sum, there has been a development that some in traditionally supportive groups—the political party in power, the electric utilities, and even manufacturers—have openly voiced their opposition or begun to exit the business.

Importantly, the NRA prevailed in early struggles and disadvantaged actors, most notably the Abe government and electric utilities, adapted to the new realities as their pressure came to nothing. In other words, initial policy entrepreneurs who had moved into the NRA to pursue their vision of independent nuclear safety governance prevailed, with continued support of a critical domestic and international audience that had already played an important role as external entrepreneurs during the reforms. Criticism from both pronuclear and antinuclear actors was a testament to the NRA's independent position between government and utilities on the one hand and antinuclear activists on the other. The new safety agency redistributed power from utilities to the public by limiting utilities' influence on regulation and by providing more opportunities for those seeking to challenge energy policy decision-making.

In response, dethroned utilities were not making the massive investments necessary to meet government targets. Those that did encountered stronger opposition than before. After 3.11 demystified Japan's safety myth, transparent safety governance broke the information monopoly previously held by pronuclear actors, and NRA safety requirements galvanized antinuclear protests. This was tantamount to losing depoliticization power, and it incurred increased social acceptance costs on top of technical safety costs. Under conditions of heightened public attention, the NRA's stringent and transparent safety governance fundamentally altered the politics of nuclear policy implementation. While traditionally pronuclear actors remained in decision-making positions, a more transparent and participatory implementation approach created a challenge for nuclear reactor restarts (Koppenborg 2021). Hence, safety administration reforms that intended to enable Japan's continued reliance on nuclear power in fact produced something more akin to a de facto nuclear phaseout.

CONCLUSIONS AND IMPLICATIONS

The starting point for this book was the puzzling existence of an independent nuclear safety regulator in Japan, the Nuclear Regulation Authority (NRA), which was able to withstand pressure to return to pre-3.11 business as usual. The analysis of Japan's regulatory reforms has shown that the addition of an independent nuclear safety regulator has fundamentally altered nuclear safety governance in Japan and, by extension, nuclear power politics.

Regulation, including safety, has become ubiquitous in modern societies. It applies to everything from everyday items, such as kitchen appliances, to more complex technologies that come with many promises and potential pitfalls, such as nuclear power and artificial intelligence. The study presented in this book is based on regulatory politics, that is, the insight that safety governance institutions and politics are not separate from each other, but rather intertwined. Regulatory governance shapes the politics surrounding a technology by organizing different interests in or out of politics. Depending on the nature of regulatory politics, the power distribution between industry, politicians, governmental administration, and the public varies in deciding—and implementing—political responses. In case of Japan, the contrast between industry-led safety governance lacking in public accountability and independent transparent safety regulation is extremely apparent. By tracing the process of creating an independent safety agency and how it redefined the relationship between state, industry, and public within Japan's nuclear policy domain, this book shines a light on the dialectic relationship between politics and safety governance.

This chapter summarizes the main empirical findings from the analysis, including both the process and outcome of post-3.11 nuclear safety governance reforms in Japan, and relates these findings to several major theoretical debates in the literature. Finally, the chapter draws some historical lessons for ongoing and future attempts to regulate established and emerging high-risk high-return technologies.

State Promotion and Regulatory Capture—until 3.11

Japan's path of developing nuclear power for half a century, until 2011, rested on a mighty pronuclear policy coalition—the so-called "nuclear village." The long-term ruling Liberal Democratic Party (LDP), the Ministry of Economy, Trade and Industry (METI), and the nuclear industry, including power plant manufacturers and regional electric utilities, jointly pursued nuclear power development in the name of energy security, affordable electricity, and environmental friendliness. They preserved the integration of nuclear power promotion and regulation, rejected independent safety regulation, and kept the public out of safety debates as much as possible. The power of pronuclear actors was buttressed by state intervention to support nuclear power and a selective information-release system that allowed pronuclear actors to influence public opinion. Furthermore, widely accepted supporting ideas, such as the priority of economic growth and energy security, bolstered by the pronuclear coalition's active promotion of a safety myth about the absolute safety of Japan's nuclear power technology, sustained public acceptance of nuclear power. Opposition consisted of NIMBY protests against the siting of individual nuclear power plants. In case citizens sought to challenge nuclear power plants via the judiciary, courts ruled in favor of nuclear power projects. Thus, pre-3.11 safety governance institutions aligned with policymaking institutions to limit participation in decision-making and implementation to a few pronuclear actors while largely keeping critics out.

While the NAIIC concluded that 3.11 resulted from "collusion between the government, the regulators and TEPCO, and the lack of governance by said parties" (NAIIC 2012, 16), the idea of regulatory capture faces criticism due to a lack of evidence of the industry actually taking over the regulator (Scalise 2015). In the case of Japan's nuclear village, capture was not the result of a takeover, but rather of a long-running intensive exchange between state and industry. Early on, the industry became a partner in Japan's nuclear power program, including policymaking and setting safety standards. This cooperative setup became flawed as accidents inside and outside of Japan revealed the risks of nuclear power, but the industry's position as a partner enabled it to block more stringent safety regulation. Largely cosmetic nuclear safety governance reforms in the 1970s and early 2000s failed to strengthen nuclear safety governance and never touched electric utilities' decision-making power over accident countermeasures and thus safety costs. As the risks involved became clear, the balance between heeding particular industry interests and improving safety at nuclear power plants via stringent regulation shifted toward the former. Mere years before 3.11, calls by the IAEA and proposals by a Nuclear Legislation Study Group to increase regulatory independence and to improve accident countermeasures fizzled out due to anticipated industry opposition. By that time, Japan's nuclear safety governance had been captured by the nuclear industry, not via a hostile takeover but rather by blocking any move to limit the nuclear industry's ability to influence safety regulation.

After the March 2011 nuclear accident at the Fukushima Daiichi nuclear power plant exposed the safety myth and shook Japan's nuclear power policy to its core, many scholars asked whether the ensuing safety administration reforms fundamentally altered nuclear safety governance and nuclear policy in Japan. Some attested to the NRA's apparent ability to resist political pressure but expressed doubts about its ability to remain independent in the long run (Aldrich 2014; Hymans 2015; Shiroyama 2015). The findings here show that nuclear safety administration reforms have, in fact, brought about fundamental lasting changes by establishing an independent, neutral, and transparent nuclear safety regulator. Fundamental changes were the result of crucial decisions that were made and institutionalized early on, in line with historical institutionalism's focus on temporality. Strikingly, the creation of an independent transparent safety agency, intended to regain domestic and international trust in Japan's nuclear safety and to thereby enable continued reliance on nuclear power, turned into a game changer for nuclear safety governance and nuclear power politics.

Nuclear Safety Reforms as a Game Changer

This begs the question of what enabled the NRA to avoid capture and to put down roots as an independent safety regulator. Studies on the rise of independent regulatory agencies (IRAs) have produced a list of organizational characteristics as well as external factors that affect an agency's independence, paying attention to both formal and de facto independence (Hanretty and Koop 2013). Formal independence pertains to the official rules regarding 1) the chairman and the management board, 2) the agency's formal relationship with elected politicians, 3) finances and internal organization, and 4) regulatory competencies (Gilardi and Maggetti 2011). For de facto independence in an agency's day-to-day interactions with the regulated interest, it is paramount for the agency to retain independent expertise that does not rest on industry input (Wilson 1989; Gilardi and Maggetti 2011). Other characteristics conducive to independence are public accountability (Jordana, Fernández-i-Marín, and Bianculli 2018), if not public participation (Beck 1992). Next to these organizational characteristics, there are certain external factors, such as audiences monitoring an organization (Carpenter 2001; Carpenter 2010), heightened public attention (Wilson 1989), and involvement in regional or international networks (Eberlein and Newman 2008; Groenleer 2012; Bianculli, Jordana, and Juanatey 2017) that can help bolster agency independence.

Beginning with formal independence, the NRA Act bestowed the new agency with the legal status of an "independent commission organization," which includes managerial autonomy, the notion of political independence, and sole regulatory authority over nuclear safety. Similar to a ministry, the NRA holds the right to draft bills, which are then submitted to the Diet (the Japanese parliament) via the minister for the environment. Board members hold full-time positions and are banned from holding additional positions in a related field, an aspect that sets the NRA apart from other commissions.

The first board, led by Chairman Tanaka, proactively utilized the bestowed powers in order to realize its independent safety governance vision. To begin with, board members successfully submitted bills in order to include transparency, staggered terms of office for board members, and the integration of yet another agency (JNES), in charge of monitoring safety at nuclear power plants, into the NRA Act. In doing so, they enshrined public accountability, continuity on the board, and monitoring on top of standard-setting competencies into the legal framework. Furthermore, within the first few months the first board adopted standards that expanded the NRA's regulatory capabilities to include sanctioning powers vis-à-vis electric utilities and passed guidelines to limit exchanges with the regulated interest and the remainder of the bureaucracy. Using far-reaching managerial autonomy granted by its legal status, Chairman Tanaka and the first NRA board solidified formal and de facto independence with amendments to the NRA Act, organizational guidelines, and new safety standards. The first board's proactive pursuit of transparency, stronger regulatory capabilities, and in-house expertise, enabled by a legal status that grants far-reaching managerial autonomy and an outgoing government supporting the young agency's efforts, was crucial to buttressing the agency's de facto independence.

Politically Independent Standard Setting

Chairman Tanaka's retirement in 2017 was a litmus test for the NRA's susceptibility to political influence. Had the government selected somebody determined
to return to pre-3.11 safety governance, it could have changed the NRA's course.
Instead, Fuketa Toyoshi, who was instrumental to the NRA's thorough safety
checks and the last remaining member from the groundbreaking first board, became the new chairman. With Fuketa succeeding Tanaka, the NRA clearly
showed continuity in leadership determined to keep a distance from politicians,
the administration promoting nuclear power, and the nuclear industry.

Not long after its election in December 2012, the LDP government under
Prime Minister Abe began pressuring the NRA. It called on the agency to water
down its safety standards, particularly the newly introduced forty-year cap on a
reactor's life span, by adopting nuclear policy goals that require reactors to be
licensed to operate for sixty years. Furthermore, a group of LDP politicians, led
by former reform leader and outspoken advocate of agency independence Shiozaki Yasuhisa, accused the NRA of "isolation" and requested the agency to give
more consideration to political concerns (Amano 2015, 21). Despite the government clearly communicating its wishes for less stringent safety standards and
more political influence, the NRA has kept the new standards in place and enforced them.

New safety standards were considerably broader in scope and more stringent
than before. This included the new life-span limit of forty years for nuclear reactors with a potential extension after additional safety refurbishments and
checks. In a major change, accident countermeasures became subject to state
regulation instead of implemented at electric utilities' discretion. The introduction of the backfit system also provided the NRA with leverage over electric utilities via threatening to revoke an operating license should they fail to comply.
Furthermore, thorough emergency planning became mandatory for a precautionary action zone (PAZ) of thirty kilometers around a nuclear power plant.
Varying in scope from area to area, new safety standards exhibit a focus on lessons learnt from the Fukushima Daiichi nuclear accident and left considerable
room for improvement regarding a cyberattack or other IT-related safety risks.

Here, a caveat is in order. The NRA's regulatory capabilities are more limited
in areas such as nuclear security and safeguarding, for example due to international treaties and commitments the Japanese government has entered. The NRA
has assumed oversight over the decommissioning process, which was largely
omitted by the NRA Act, but holds no sanctioning powers in this area. Hence,
the NRA holds full regulatory capabilities with regard to nuclear safety, but not
nuclear security, safeguards, and decommissioning.

Furthermore, regulatory independence and conducting the most stringent safety assessments are two different animals. This goes back to a long-standing debate about the appropriate nature of risk assessments. While the precautionary principle postulates that a product or technology is presumed harmful until proven safe, the traditional risk regulation approach assumes the same technology to be safe until proven harmful. This often results in battles over the exact meaning of "due caution" (Durant 2004), especially in cases where science cannot fully establish the nature and magnitude of risks. This is the case with the risks posed by earthquakes and volcanic eruptions to nuclear reactors in Japan, both of which are difficult to predict exactly. For related risk assessments, the NRA made probabilistic hazard analysis mandatory, rather than the more sophisticated probabilistic risk assessment, a global benchmark for safety evaluations. During safety screenings, the NRA accepted utilities' risk assessments, which included proven sources of risk but excluded other potential hazard sources. Considering that scientific agreement enhances scientific authority, and a lack of scientific consensus opens up room for political debate and maneuvering (Scalise 2015), the NRA has adopted what could be called the more politically safe risk assessment approach. In doing so, the NRA rejected a precautionary approach, but implemented rigid regulation within the bounds of conventional risk regulation.

Securing Expertise

The importance of independent expertise was clearly recognized by the NRA, whose mission statement stresses avoiding regulatory capture as a result of lacking expertise within the regulator (NRA 2013b). This was rooted in the NAIIC's insights that its predecessor's flaws also stemmed from officials later retiring to the industry they used to oversee (*amakudari*), with regular staff turnover in the regulating agency limiting the role of specialists while simultaneously increasing the need for industry experts to temporarily work as regulators. With the establishment of the NRA came, for the first time, the possibility for public officials to embark on a career as nuclear regulators. For this purpose, all NRA staff, including newly hired graduates, midcareer experts, and officials taken over from its predecessors, had to undergo in-house training. With help from abroad, especially the US NRC, the NRA's Human Resource Development Center devised increasingly sophisticated training programs, which came to include three different career paths and license examinations for safety inspectors. Furthermore, guidelines on the neutrality of external expertise largely excluded experts with close (financial) ties to the industry from safety discussions. The NRA further removed itself from common practices in the Japanese administration, such as *amakudari*, and limited staff exchanges with other parts of the bureaucracy to avoid conflicts of interest between nuclear regulation and promotion. In sum, the new safety agency went to great lengths to raise its in-house expertise and to ensure that a lack of expertise on its part would not become another gateway for influence by the industry and nuclear power promoters.

An important step was to limit staff exchanges with nuclear power promoters in the Japanese bureaucracy, most notably METI, which closed a door for indirect political and industry influence. The focus of analytical schemes, however, is on relations between politicians and regulatory agencies (Gilardi and Magetti 2011, Jordana, Fernández-i-Marín, and Bianculli 2018), which mirrors the long-standing focus of debate in political science on their respective roles in governing modern states. As Guidi, Guardiancich, and Levi-Faur (2020) point out, there is a need for studies on regulation to pay greater attention to the political economy. The analysis of Japan's independent nuclear safety agency highlights that it is important to pay close attention to the bureaucracy in so called developmental states with a coordinated political economy, and more generally states with a strong bureaucracy.

Importance of Transparency

Transparency was an aspect the first board pursued from the very start as a means to regain domestic and international trust in Japan's safety regulation. Guidelines for transparency and the disclosure of administrative documents interlinked to form a comprehensive information release system not subject to requests. Most meetings were not only streamed live on the agency's YouTube channel "NRAJapan" and the Japanese platform NicoNico but were also open to people wishing to attend in person. The NRA further opened its press conferences to everyone, including civil society representatives. It thereby renounced the "journalist club" (*kisha club*) system used by the Japanese administration and criticized for contributing to a pro-elite bias in media reporting (Freeman 2000). Far-reaching transparency increased public accountability, including about the residual risks of nuclear power generation, and thwarted any government attempts to create another safety myth, this time based on the notion of stringent regulation. The new agency also published draft guidelines, regulations, and bills for public comments before making a final decision on them. In response to demands for more public participation, there was an increasing number of hearings in local communities seeking restarts. Transparency served to keep public opinion informed, which significantly improved public accountability. Vice versa, the public watching regulators live at work—online or as observers at board meetings—had the effect of keeping regulators on their toes. Furthermore, transparency protected the agency from overt attempts to undermine its independence.

A caveat is that the amount of transparency the NRA can provide regarding nuclear security and safeguards is bound by security concerns related to the misuse of nuclear materials. The State Secrecy Law, passed by the Abe government in 2013 and the impetus for massive public protests, allows the government to designate defense-related and other sensitive information as so-called special interests that may not be disclosed. This nondisclosure requirement applies to certain NRA meetings that discuss nuclear security issues and the safeguarding of nuclear materials. Hence, the NRA provides full transparency with regard to nuclear safety but not nuclear security and safeguards. Despite these limitations in some areas, the NRA's transparency increased public accountability and opened the door to external scrutiny—and thereby buttressed regulatory independence.

Audiences and Internationalization

The overarching reform aim to regain the domestic and international trust lost after 3.11 made the Japanese public and international actors an important audience. This was also evident in the NRA's internationalization efforts. Next to actively cooperating with international peers, it invited international scrutiny in the form of the IAEA Integrated Regulatory Review Service (IRRS) evaluating the NRA's progress toward becoming an independent regulatory agency. With many recognizing the IAEA as an international authority on safety governance, IAEA evaluations lend legitimacy to the NRA's independent approach. Vice versa, had the IAEA, or the US NRC for that matter, expressed doubts about the independence of the NRA, this would have given the Japanese public great reason to doubt the veracity of government statements about strong safety regulation. As IAEA standards came to include "freedom from political pressure" (IAEA 2016a) in response to the lessons learned from 3.11, IAEA scrutiny also helped ward off pressure from pronuclear actors. Furthermore, IAEA recommendations helped legitimize NRA actions unpopular with operators, such as more intense and critical safety inspections, putting an end to an inspection system that was more akin to utilities giving regulators a guided tour. As such, active engagement with international peers through the IAEA buttressed the NRA's independence.

For the emergence of the NRA as an independent nuclear safety agency in Japan, transparency—and the domestic and international scrutiny it invited—was crucial. This reconfirms previous findings about the importance of audiences monitoring an agency (Carpenter 2001; Carpenter 2010; Wilson 1989). A current research theme in the study of regulatory agencies is to elucidate how interactions with different audiences and networks shape agency independence (Maggetti and Verhoest 2014). The NRA's case lends further evidence to the concept of "international networks as drivers of agency independence" (Bianculli, Jordana, and

Juanatey 2017), but adds certain context conditions. The first condition is public attention in combination with politicians attributing importance to the notion of an independent regulator. Neither was the case in Japan prior to 3.11, when IAEA criticism of Japan's nuclear safety administration and recommendations for greater independence came to nothing. After 3.11, however, the government stressed the notion of an independent technocratic regulator, a condition under which active involvement in international networks, particularly IAEA assessments, has buttressed agency independence. The second condition is related to an agency's formal independence. If it enjoys sufficient managerial autonomy to implement IAEA recommendations for greater independence regardless of domestic audiences, the first context condition becomes less of a constraining factor. Clearly, international networks are important, because they attribute importance by definition while domestic audiences are more fickle.

Independent Regulators in Japan

The level of independence achieved by the NRA is unusual for Japan, where policy-making, promotion, and regulation tend to be located inside the same ministry. Another example, the Japan Fair Trade Commission (JFTC), was "notoriously weak" during the first decades of its existence (Vogel 2006) but strengthened in the early 2000s, making it a "rare independent regulator in Japan" (Mogaki 2019, 17). The NRA and the JFTC share a few commonalities. Both were established with the legal status as an independent regulator based on Article 3 of the National Government Organization Act, which helped shield the JFTC against external influence, especially from METI (Mogaki 2019), as it did for the NRA. Furthermore, the United States has played a significant role in the establishment of both. The JFTC was established shortly after the end of World War II under the US occupation of Japan, while the US NRC and its independence served as a model for post-3.11 nuclear safety reforms. However, there are other regulatory institutions based on Article 3 that are far less independent, such as the Japan Transport Safety Board and the Environmental Dispute Coordination Commission, pointing to the need for further research about the factors that enable regulatory agencies in Japan to translate their legal status into independence in its day-to-day operations.

Potential Warning Signs

From the analysis results, we can derive a number of warning signs that would indicate the NRA backsliding on its independence. An obvious sign would be the government making changes to the NRA Act in such a way that restricts the NRA's managerial autonomy or any of the amendments the first board has made to the law. Obviously, budget and resulting staff cuts to the agency, but also to its Human Resource Development Center, would impede its ability to rely on independent expertise. Given far-reaching managerial autonomy and the powerful position of the chairman, the appointment of an agency head determined to water down safety standards and to heed political and industry concerns would be another red flag. As this book goes into production in September 2022, the term of the second NRA Chairman, Fuketa, just came to an end. His departure marked the last inaugural board member leaving almost exactly ten years after the agency's inauguration in September 2012. The new chairman Yamanaka Shinsuke has been a board member since 2017 and pledges to continue "nuclear safety regulations based on independence and transparency" (NRA 2022). It is too early to tell what kind of chairman he will be, but it is a promising sign that someone from within the board was elected rather than a candidate newly imposed by the government.

It is not as clear how a more positive public attitude to nuclear power would immediately impact the NRA. Even though public opposition to restarts dropped to 47 percent in February 2022 (Asahi Shimbun 2022), critics challenging NRA decisions via lawsuits will most likely continue to do so. However, a shift in public opinion could embolden the government to pressure the NRA to speed up reactor restarts and to water down safety standards. In fact, in August 2022 Prime Minister Kishida Fumio called for reactor restarts and suggested that a government committee look into ways to extend the reactor lifespan. This took place against the background of soaring prices and a gas shortage due to the Russian invasion of Ukraine. However, he also qualified his call to apply only to reactors that had already received green light by the NRA (Reuters 2022) but were offline due to court orders among other reasons. The suggestion to look into a lifetime expansion was too vague to tell what it would mean exactly. Depending on what the government resolves to undertake, it could amount to a challenge of the 40-year limit introduced by the NRA—or something entirely different, such as more financial support for utilities seeking to refurbish existing reactors in order to receive an operating license beyond 40 years.

As the NRA celebrates its first decade in existence as an independent and transparent safety agency, it is undergoing a change in leadership and is facing potential challenges, such as a government looking to increase the share of electricity produced by nuclear power and a slight shift in public opinion toward a more favorable position on nuclear power. Domestic pressure could potentially be offset by sustained international scrutiny by the IAEA and international peers, especially the US NRC. Should both domestic and international scrutiny wane, it would leave the NRA in a more vulnerable position but might not have an immediate impact on its well-established independence. At the time being, the most decisive factors appear to be the new chairman's dedication to upholding independence and transparency and the lengths to which the Kishida government is willing to go in order to bring nuclear reactors back into operation.

Should the NRA continue on its current trajectory, it will become somewhat weaker in the longer run. As of early 2022, more nuclear power plants were in the process of being decommissioned than in operation. This includes damaged reactors such as those at the Fukushima Daiichi nuclear power plant. Regulatory capabilities are lower for decommissioning than for the restart process. Unless the NRA finds ways to increase its regulatory competencies in this area in a way similar to reactor safety, as the balance shifts toward overseeing decommissioning efforts, the NRA will become weaker in terms of its regulatory competencies.

Post-3.11 Politics of Regulatory Design

These developments in Japan's nuclear safety governance raise questions about how the nuclear village, and utilities in particular, lost out during the reform process. A current research aim in historical institutionalist research focuses on better understanding mechanisms for change. Scholars of critical junctures seek to understand why some external shocks are followed by the reemergence of the same institutional arrangements, termed a "near miss," while others result in fundamental changes and therefore serve as a critical juncture (Capoccia 2016a). This is tied to a debate about ideas and power-based explanations for why actors opt for a certain institutional design while disregarding others (Béland, Carstensen, and Seabrooke 2016; Capoccia 2016a; Hogan 2019). Both explanations draw on the notion of previously uninvolved actors becoming agents of change that push for reforms. The case comparison conducted here, between an aborted attempt to institute a nuclear phaseout policy and fundamental reforms of the nuclear safety administration, is well suited to discern the importance of agents of change, their respective reform ideas, and power in the reform process.

In 2011, the initial reform impetus came from the Democratic Party of Japan (DPJ), a political entrepreneur that had been the main opposition party before coming into government for the first time in 2009. For one, the DPJ government sought to identify ways for Japan to phase out nuclear power. In the process, it consulted the Japanese public via new participation tools such as deliberative polling. Despite public pressure for a nuclear phaseout, the DPJ government failed to push through such a policy. Plagued with internal divisions on the matter, DPJ governments under Kan and Noda were unable to adopt a Cabinet

Decision that would put the issue on the legislative agenda. Faced with severe opposition from business interests and in the absence of international pressure, the DPJ government produced only a vague resolution to "reduce dependence" on nuclear power. Once the LDP won the 2012 general elections, energy policy-making fell back into the hands of traditional power elites and supporters of nuclear power, who have declared retaining a 20–22 percent share of nuclear power by 2030 as the extent of reductions possible.

Meanwhile, the DPJ Cabinet adopted a Cabinet Decision in August 2011 on reforming Japan's nuclear safety administration. Using its power in government, the DPJ Cabinet structured the reform process in a manner that kept old elites out as much as possible and preempted their veto power, most notably METI and nuclear industry actors that were instrumental to previous reforms resulting in mere window dressing. Instead, it brought previously sidelined domestic and international experts in as policy entrepreneurs. In 2011 and 2012, more agents of change joined the choir of voices calling for the creation of an independent nuclear safety agency, including demonstrators, domestic and international experts, and the LDP-led opposition. With the necessary votes from the opposition, the NRA Establishment Act passed the Diet and laid the legal foundation for fundamentally different governance institutions. Hence, the DPJ government and other agents of change eager to improve nuclear safety governance consolidated around the idea of an independent safety agency and successfully pushed for reforms.

In the safety reform process, three decisions were of particular importance. The first one was the Cabinet Decision on the separation of nuclear power promotion and regulation, which stipulated a safety agency with stronger in-house expertise and regulatory competencies that would be located outside of METI. The ensuing debate in the Diet revolved around the appropriate legal status of the new agency. While the DPJ advocated for an agency closely connected to the Ministry of Environment, but with managerial autonomy, the LDP attacked this proposal as being out of touch with IAEA standards for independent safety agencies. As opposition votes were necessary to pass a bill, the LDP was able to push through not only its demand for a strong legal independence based on Article 3 of the National Government Organization Act, but also the removal of concrete stipulations that would enable the new regulator to act independently in practice. Once the NRA Act had been passed, the DPJ Cabinet gave designated board members and senior NRA officials free rein in designing the new agency. They developed many of the rules and guidelines that subsequently empowered the safety agency to defy pressure from pronuclear actors and to remain independent. The cabinet also helped them enshrine their vision into the NRA Act via supplementary provisions, which included provisions the LDP had blocked in the process, such as top-down safety standards. Hence, the Noda Cabinet as a political entrepreneur and designated NRA officials as policy entrepreneurs finalized nuclear safety governance reforms that would break with Japan's previous path of industry influence—just in time before the LDP returned to government and began pressuring the new agency to heed its nuclear policy concerns.

Political Leadership by the DPJ

The DPJ swept into power in 2009 with a reformist agenda that promised to overhaul decision-making systems in order to establish "politician-led government" (*seiji shudō*) with a strong prime minister and politicians prevailing over bureaucrats. However, the DPJ failed to deliver on the promised overhaul of Japan's decision-making systems (Kushida and Lipscy 2013a; Zakowski 2015; A. Shiozaki 2017). While the DPJ may have failed to change Japan's decision-making institutions as such, it exhibited leadership in both the nuclear safety administration reform and the phase-out attempt by holding reform deliberations at the Cabinet Office rather than relying on ministry-led policy deliberation councils. In the case of safety governance reforms, somewhat ironically, political leadership by the DPJ produced a regulatory agency far removed from political influence.

Ideas and Power as Change Mechanisms

Returning to the debate about ideas and power-based explanations for why actors opt for a certain institutional design while disregarding others (Béland, Carstensen, and Seabrooke 2016; Capoccia 2016a; Hogan 2019), the findings suggest that ideas and power as change mechanisms interact during a critical juncture. While the relationship between the two mechanisms for change has remained understudied, clearly political entrepreneurs are at the nexus of policy ideas and political power (Béland, Carstensen, and Seabrooke 2016).

Concretely, this research identified two ways in which political entrepreneurs are at the nexus of policy ideas and power. For one, political entrepreneurs' power to structure the process can help agents with new ideas enter the process while keeping those vested in the discredited policy or institution out of reform efforts. This was the case in both the aborted nuclear phaseout and the successful safety governance reforms. Second, and more importantly, agreement among political entrepreneurs is what bridges power and ideas. The reason the DPJ Cabinet under Kan was able to set the reform trajectory early on was its united stance on the creation of an independent safety agency as stipulated in the Cabinet Decision. While further support was necessary, especially from the LDP, in order to pass a reform bill, this early unity as well as the later consensus on supporting designated agency officials in bringing their independent governance vision to life was crucial. Generally speaking, political entrepreneurs are collective actors in potentially powerful positions. A precondition for wielding power is agreement on which decision to take. Put differently, consensus on an alternative idea is a precondition for political entrepreneurs to exercise their power to overcome resistance by those vested in and empowered by the extant institutional order. Hence, the nexus pointed out by Béland, Carstensen, and Seabrooke (2016) can be thought of as a two-step process: First, there needs to be consensus on an alternative idea among political entrepreneurs, which, second, puts them in the position to act more decisively to overcome vested interests.

Political Will

The nuclear accident at the Fukushima Daiichi plant was, as a shock to Japan's system, an important context condition for fundamental nuclear safety governance reforms. Despite "a common tendency [. . .] to assume that failures occurring in other countries are unlikely to be repeated in one's own" (Sagan 2016, 6), a potential way out of the improvements-in-response-to-accidents-and-scandals dynamic would be to learn more from regulatory failures and improvements within the same sector in other countries. Here, the rise in regulatory agencies can be regarded as an indicator of political will to make a more credible safety and risk-reduction commitment to citizens. An understanding of the interlinkages between regulatory governance and politics can point out patterns in safety governance and make researchers and practitioners aware of aspects that determine where safety governance ends up on the spectrum between private-interest and public-interest regulation.

A first lesson to draw is the importance of political will to establish an independent regulatory institution. The process of creating Japan's independent and transparent nuclear safety agency was a testament to the politics of regulatory agency design. In the design process, the political power constellation and interests at the time were a decisive factor for where safety regulation would fall in the spectrum of balancing public and industry interests. The degree of independence attained by the NRA resulted from the will of political decision makers and designated regulators to design an independent and transparent agency that would be able to stand its ground. Considering findings about the weakness of regulating agencies responsible for risk and social areas, especially nuclear safety, when compared to telecommunications and banking for example (Jordana, Fernández-i-Marín, and Bianculli 2018), an interesting avenue for further research would be do delve into the political reasons and incentive structures, across countries and political systems, that have led decision makers to design less independent agencies in social and risk areas, despite their potentially large impact on public health and safety.

Impact on Japan's Nuclear Village

Adding an independent safety regulator altered interactions within the nuclear village as well as between pronuclear actors, antinuclear activists, and the public. A policy domain is structured by institutions granting access to decision-making to some while excluding others. Those in decision-making positions, the winners of an institutional order, constitute the "policy monopoly" (Baumgartner and Jones 1993) or "dominant coalition" (Sabatier and Weible 2007). For the winners of an institutional order, limiting the number of actors is an important strategy in politics, mainly by keeping power-buttressing institutions and supporting policy ideas in place. In the pre-3.11 nuclear energy policy domain, pronuclear actors successfully limited public participation, shaped public opinion, and devised ways around local protest in order to develop Japan's nuclear power program. With the LDP's reelection, the old power elites were once again in charge of devising nuclear power policy goals, but the NRA's safety governance targeted the feedback mechanisms that had previously buttressed pronuclear actors' power to implement these policy goals. Hence, the reformed safety regulation agency redistributed power among different actors in the nuclear policy domain and thereby reshaped Japan's nuclear energy policy conflict.

Governance by an Independent Regulator

Within the nuclear triangle, much of the technical expertise and policy implementation power previously resided with electric utilities, putting them in a powerful position. Nuclear safety reforms, however, created a fundamentally different business environment for electric utilities. The NRA imposing binding safety regulation removed utilities' decision-making power over safety regulation, particularly accident countermeasures and emergency planning. This shifted the political economy away from "profitable and virtually risk-free" nuclear power expansion (Cohen, McCubbins, and Rosenbluth 1995, 193) based on "reciprocal consent" (Samuels 1987) in the heyday of Japan's nuclear power development. Playing on the quote from Samuels that "the Japanese bureaucracy does not dominate, it negotiates" (1987, 260), the NRA does not negotiate, rather it dominates nuclear safety—to the dismay of the nuclear industry.

This has implications for the question of who governs Japan. As a consequence of adding an independent regulatory agency to the nuclear village, the power dynamics shifted. Starting with elements of continuity, the Abe government continued the legacy of previous LDP governments that mainly delegated the task of nuclear energy policymaking and implementation to METI. However, METI lost competencies related to licensing and nuclear safety to the NRA. Interactions between the NRA and the nuclear industry have come to closely resemble a "regulatory state" (Levi-Faur 2011), which relocated power from the industry to a state regulator. In line with the finding that business interests win during times of "quiet politics" (Culpepper 2010), the nuclear industry lost its position of power within the pronuclear coalition after the March 2011 nuclear accident politicized nuclear power. However, neither the LDP as the principal (Ramseyer and Rosenbluth 1993) nor METI as a powerful ministry (Johnson 1982; Pempel 1982) gained power in nuclear safety regulation. Rather, an independent agency, an unusual feature in Japanese governance (Mogaki 2019), has taken over nuclear safety governance. It remains to be seen whether the NRA remains the exception or whether Japan will join other countries on their path toward establishing more independent regulatory agencies.

Expanded Nuclear Policy Conflict

Independent nuclear safety governance also had an impact on nuclear power opposition. While large-scale antinuclear protests against nuclear power subsided after 2012, new safety regulation galvanized local civil society opposition. For one, expanded emergency preparedness regulation, particularly the thirty-kilometer precautionary action zone around nuclear power plants, affected previously uninvolved municipalities across prefectural borders. As a result, district courts in adjacent prefectures gained jurisdiction, which increased the number of available venues for local civil society opposition via lawsuits. In other words, the NRA—unintentionally—broadened the conflict scope by drawing more communities into the struggle over restarting nuclear power plants. Furthermore, new safety requirements, such as a second license after forty years in operation, provided citizens with more opportunities to challenge a reactor's operating license. Many lawsuits challenged the NRA's conservative risk regulation approach, thereby demanding a more precautionary approach instead. As a result of an expanded metaphorical backyard and more opportunities to challenge reactor restarts, nuclear power opposition evolved from NIMBY "site fights" (Aldrich 2010) into cross-prefectural reactor fights centered on due caution.

The resulting surge in class-action lawsuits against nuclear power plants has strong implications regarding the future of nuclear power in Japan because courts are the only bodies with the authority to override NRA decisions. Japan's courts can function as an avenue either for citizens to engage in politics, as was the case with stronger pollution regulation (Upham 1987), or for politicians to keep them out via court decisions, as seen with nuclear power plant cases before 3.11 (Kingston 2014; interview with H. Ban, 2018). The tendency of higher courts to rule in favor of the LDP's stance on high-profile partisan disputes has been attributed to the government using the appointment of judges as a way to punish those who ruled against its position (Ramseyer 2019). While the majority of post-3.11 class-action lawsuits were still pending as of September 2022, court decisions already hint at a greater receptiveness of the judiciary for the safety concerns raised by citizens. Court rulings in favor of nuclear power opponents questioned the LDP's hold over the judiciary and posed an additional restart hurdle to utilities that had made enormous investments into safety refurbishments.

Pre-3.11 path dependency rested on low technical and social acceptance costs and was further reinforced by institutional arrangements in the energy domain (energy monopolies), academia (close industry ties through funding), Japanese media (journalist clubs and advertisement money), and the judiciary (political influence via judiciary appointments). The first, energy monopolies, was actually abolished by the LDP as part of its energy market liberalization policy after 3.11. Second, nuclear experts with close ties to the industry, financial and otherwise, were mostly excluded from NRA nuclear safety deliberations. Furthermore, the NRA's transparent approach has renounced the system of journalist clubs and has kept public opinion informed about nuclear safety. Stressing remaining risks has thwarted government attempts to create a new safety myth based on the idea of the NRA certifying safety. The fact that mainstream media reporting on nuclear power has become more polarized and includes more diverse opinions (Weiß 2019) indicates a lessened influence of the nuclear industry via advertising money. Lastly, the judiciary has become embroiled in the conflict over nuclear restarts rather than acting as a reliable gatekeeper keeping protestors away from nuclear power. Hence, previously reinforcing institutional arrangements have ceased to apply to nuclear safety.

Coming back to the pronuclear coalition, NRA safety governance undermined the feedback mechanism that previously glued the dominant coalition together. Pre-3.11 safety governance institutions aligned with policymaking institutions and limited participation in decision-making and implementation to a few pronuclear actors while largely keeping critics out. In the language of costs often used by historical institutionalist scholars, the nuclear village kept technical costs down at the expense of public safety and limited social acceptance costs by keeping the public and critics out. Under conditions of regulatory capture and limited public participation in the safety debate, both of which helped keep the costs of nuclear power low, Japan's nuclear iron triangle pulled in one direction. Even before 3.11, however, costs emerged as a possible Achilles' heel of the system, and what appeared to be a strong METI-centered nuclear power program during the 2000s was actually an attempt to revive stagnating development. The NRA marked a critical juncture in Japan's nuclear safety governance. It redistributed nuclear safety implementation power from industry to regulator, expanded the number of participants in the policy implementation conflict, and prevented the resurgence of a nuclear safety myth. As new safety standards led to an explosion of technical costs and a broadened conflict scope incurred further social acceptance costs, this new situation weakened the power structures, the glue, that had bound nuclear proponents together.

A Critical Juncture for Nuclear Policy Implementation

Seemingly small changes in Japan's nuclear safety governance, initially designed to enable a continued use of nuclear power, have in fact ushered in a new path. Independent and transparent safety governance undermined the feedback mechanisms that previously sustained the joint stance of the dominant coalition and its power to implement nuclear policy and, by extension, energy and climate policy (discussed in more detail in the next section). Hence, this book argues that even small reforms can become political game changers, if they weaken the position of powerful actors that sustained the previous policy path. These findings point to the need to specify which institutions and underlying feedback mechanisms are crucial. Building on the argument by Capoccia and Kelemen (2007) and Capoccia (2016a) on the need to clearly identify institutional change with regard to a specific unit of analysis, the case of Japan highlights the importance of specifying relevant institutions and feedback mechanisms in a given policy domain for both policymaking and implementation.

Fissures in the Nuclear Village

Safety governance reforms created fissures in Japan's previously dominant pro-nuclear policy coalition, the nuclear village, for example, between the LDP and electric utilities. To begin with, the LDP joined the choir of voices blaming TEPCO, and by extension all electric utilities, for the nuclear accident but refrained from apologizing from its own involvement in the "collusion" decried by the NAIIC report (NAIIC 2012). The LDP making electric utilities the sole scapegoat to defect blame from itself was not appreciated by plant operators. After the party's reelection, the LDP government under Abe failed to show the strong political support the nuclear industry was looking for. For example, the government abandoned any calls for the construction of new nuclear power plants. To make matters worse for utilities facing massive additional investments into safety refurbishments, the Abe government adopted a market liberalization in 2015 that put an end to the regional monopolies long held by electric utilities and exposed them to more market competition.

Given the LDP's long-standing relationship with utilities in developing Japan's nuclear power program, one has to wonder about the reasons behind its lackluster support for the nuclear industry, especially utilities. While still clearly pronuclear during the 2012 election campaign (Endo, Pekkanen, and Reed 2013), the LDP avoided taking a clear stance ahead of the 2014 elections (Maeda 2015). Essentially, the Abe government was faced with a choice after early pressure on the NRA came to nothing: to support the nuclear industry, and to risk nuclear power emerging as an electoral issue, or to focus on winning electoral support despite holding on to a commitment to unpopular nuclear power. To achieve the latter, the LDP under Abe sought to depoliticize nuclear power ahead of the 2014 elections by shifting responsibility for restart decisions to the NRA (Hughes 2015). This was clearly reflected in the language used in Strategic Energy Plans since 2014 and by the government stressing that restarts would take place once the NRA deemed reactors safe. This depoliticization strategy required a credible commitment to nuclear safety, which forbade curtailing the NRA's independence and thereby limited the government's ability to support the industry by overtly pressuring the safety agency. Another support strategy, to pass utilities' additional safety costs on to consumers in the form of higher electricity prices, would have been very unpopular with the electorate given Japan's already high electricity prices. In the past, the LDP had often reacted to threats to its preeminence by extending income compensation, entitlements, and subsidies (Calder 1988), but this time there was no compensation for electric utilities. After 2012, the LDP opted for half-hearted nuclear industry support, rather than to risk its unpopular pronuclear policy emerging as an electoral issue.

But then why did successive LDP governments hold on to nuclear power? Changing nuclear power policy has direct implications for policy areas where nuclear power plays an important role. A return to a nuclear policy was considered pivotal to Abe's economic policies to reignite growth (Kingston 2016; Incerti and Lipscy 2018); to Japan's efforts to raise its energy self-sufficiency rate, which reached a low point after 3.11 (Vivoda 2014); and to lowering greenhouse gas emissions in line with international commitments under the United Nations climate regime (Kameyama 2019). Given Japan's large stockpile of plutonium, a nuclear phaseout would also have foreign policy implications. The United States expressed concerns that Japan abandoning nuclear power and the nuclear fuel cycle would raise "uncomfortable questions about the consistency of US nuclear non-proliferation efforts targeting Iran and North Korea" (Kingston 2014, 116). Just as prior to 3.11, pronuclear actors pursued nuclear power as a means to realize the 3Es of Japan's energy policy aims, namely energy security, economic efficiency, and environmental friendliness. It appears that LDP governments sought to reap the political benefits of retaining nuclear power as part of its energy, economic, and climate policy while imposing the burden of additional costs on electric utilities.

Electric utilities' response to rising technical and social acceptance costs amid lackluster political support questions the notion of a nuclear village united behind the promotion of nuclear power. As costs exploded, utilities adopted a more business-oriented approach with strategic investments only into younger and larger reactors that might still prove profitable for them. As a result, Japan saw a drastic reduction in nuclear power generation capacity. Electric utilities have applied for safety checks at twenty-seven nuclear reactors, which could provide about 14 percent of Japan's electricity mix by 2030—that is, if all of them are restarted and electricity demand remains stable at 2011 levels. There was only one further application for safety screenings between 2015 and 2022. Importantly, every year that passes removes another year off a reactor's life span, making it less likely for safety investments to be worthwhile. Assuming utilities do refurbish and restart all thirty-five remaining commercial nuclear reactors in Japan, it would amount to an 18 percent share in 2030 and, therefore, fall flat of the government target of 20–22 percent. Since utilities ceased to make the investments necessary to achieve government energy policy targets after losing their position as partners in implementing safety measures, it appears that the system of "private implementation of public policy" (*kokusakuminei*) (Kikkawa 2012) has come to an end. With electric utilities not moving to make the necessary investments to meet government targets, their actions amount to a beginning of an exodus from the pronuclear policy coalition.

There are other examples of a weakening pronuclear coalition. Prominent LDP politicians, such as former prime minister Koizumi Junichirō and former foreign and defense minister Kōno Tarō, have spoken out against nuclear power. One of Japan's nuclear power plant manufacturers, Toshiba, indicated a move away from the nuclear power business by selling its stakes in the US nuclear power business Westinghouse in 2018. At the same time, Japan's largest business federation, Keidanren, has openly criticized the lack of government support for nuclear power while utilities have done so behind closed doors. As some in traditionally supportive groups—the political party in power, the electric utilities, and even manufacturers—openly voice criticism of nuclear power or each other and others begin to exit the business, the blurring of the edges of the nuclear village makes for an interesting area for further research.

Implications for Japan's Energy and Climate Policy

The LDP retaining nuclear power as a policy pillar without the ability to achieve the targets set (Koppenborg 2021) poses a challenge to Japan's energy and climate policy. Nuclear power remains central to Japan's plans to achieve carbon neutrality by 2050, announced by Prime Minister Suga in 2020, and its "leadership" ambition (Suga 2021) in the global fight against climate change. The sluggish return to nuclear power challenges these ambitions, especially as the gap in electricity supply is mostly filled with fossil energy sources. Furthermore, the Russian invasion of Ukraine has highlighted the risks of Japan's dependence on fossil fuel imports as the rise in global energy prices let domestic energy prices soar. A potential solution to these challenges would be a swift and large-scale expansion of Japan's renewable energy generation capacity as a domestic and low-carbon energy source to replace nuclear power. While the 2015 electricity market liberalization and the higher share projected for renewable energy in Japan's electricity mix by 2030, raised to 36–38 percent in the 2021 SEP (METI 2021), were steps in that direction, they were not sufficient, yet, to put Japan's energy and climate policy back on track. Hence, the government's unwillingness to lend the support necessary to restart sufficient nuclear power plants—or to wholeheartedly embrace renewable energy expansion as an alternative—has left Japan's energy and climate policy somewhat in a state of disarray more than a decade after 3.11.

Lessons for High-Risk High-Return Technology Governance

The Fukushima nuclear accident affected nuclear power worldwide. Even before 3.11, rising costs of nuclear power were an issue beyond Japan. Looking at escalating costs in France and the United States, Lévêque (2015, 44) pointed out that "a technology with rising costs is a very strange beast, which requires closer study, particularly as this feature distinguishes it from several competing technologies, such as wind and solar." Hence, even before 3.11, the three countries with the most nuclear power plants struggled with rising costs. After 3.11, the global nuclear industry suffered from increasing costs and their negative impact on the economics of nuclear power as well as safety concerns leading to public opposition. Global nuclear power generation first dropped by 4 percent in 2011, followed by an even larger reduction by 7 percent in 2012. With many nuclear power construction projects cancelled in the aftermath of 3.11, China remains the only country with a major new-build program (Schneider and Froggatt 2013, 6; Worldwatch Institute 2013). In cost calculations of nuclear power, safety regulation is often omitted, because it is hard to measure (Lévêque 2015), and the findings here show that even Japanese utilities struggled to correctly estimate the additional investments necessary to meet new safety standards. While difficult to measure exactly, rising technical safety costs and social acceptance costs pose a challenge to the nuclear industry. This is particularly so in democratic countries, where technology choices tend to follow market mechanisms and public opinion and participation matters. As the dynamic of rising costs has a negative impact on nuclear power's competitiveness with other energy sources, this may even prove to exert some influence in authoritarian countries.

Over a decade after the devastating Fukushima nuclear accident shattered trust in nuclear safety and spawned game-changing reforms, this book seeks to distill lessons for the prevention of a low-probability but high-impact event. Many new technologies, such as artificial intelligence and its applications in smart green solutions and omnipresent digitalization, have much in common with nuclear power in terms of the promises they entail, but also the massive potential negative consequences. Just as nuclear power, these technologies are not positive in and of themselves, but rather the extent of their benefit depends on the humanly devised governance of these technologies. The high-risk high-return technologies mentioned here pose challenges for political decision makers now that are similar to those Japan faced in the 1950s when it set up nuclear safety governance. Two broader lessons for regulatory governance shall be highlighted in this concluding section: the importance of knowledge and avoiding the pitfall of technology optimism hampering effective safety regulation.

Knowledge Is Power

Knowledge matters. Concretely, the distribution of knowledge, both as technical/ scientific expertise and public knowledge, has an impact on safety regulation. First, independent expertise is an important prerequisite for state regulation. The lower the expertise on the part of governing bodies, the more dependent they are on industry expertise, which opens the door to influence and eventually capture. Hence, it is paramount that states foster the buildup of knowledge about a technology outside of the industry developing and promoting it, which highlights the issue of knowledge production and the state of research policy in emerging and rapidly developing fields, such as artificial intelligence and digitalization. Second, transparency functions as a door opener for public accountability and a societal debate about acceptable risks. Governance of high-risk technology deserves a societal debate, rather than being left to engineers and experts paid by the very same companies that profit from the technology in question. In short, the more expertise state institutions possess and the more that public accountability enables public debate, the more likely the pendulum of power distribution between state, public, and industry is to swing toward the first two.

Blind Technology Optimism

A second lesson is that technology optimism may hamper effective safety regulation. The early technology optimism associated with nuclear power, that is, providing an endless clean affordable power supply, later turned into a dangerous safety myth that defied opportunities to learn about the potential downsides of nuclear power until the March 2011 nuclear accident cast a harsh light on them. The promises newly emerged high-risk high-return technologies entail may make decision makers and the public initially blind to risks along the lines of the so-called halo effect, which describes "the common bias that plays a large role in shaping our view of people and situations" (Kahneman 2012, 82). To counterbalance the blinding halo effect surrounding new technologies, some aspects introduced in Japan after 3.11 can serve as lessons to improve safety governance in other areas and in other countries in order to prevent a similar shock event. First, transparency is crucial to enable a public debate. Second, it is important to face critics and to consider their concerns instead of silencing them, be it within the safety agency, the broader administration, academia, or society more generally. Third, state agencies require the necessary expertise to have the ability to understand the risks involved and regulate in the public interest. These three aspects—including citizens, fostering state expertise, and facing the concerns of opponents and critics—are important steps toward a publicly accountable risk debate that can help overcome the halo effect and thereby facilitate learning about risks before a low-probability but high-impact event demonstrates them all too vividly.

In conclusion, this book has underlined that technology regulation will always exist between the potentially conflicting priorities of politicians, industry, and public. The model of regulatory politics deepens our understanding of the political nature of safety governance institutions, and it offers a dynamic account of how such institutions are created and, in turn, influence levels of public accountability and industry influence over the risks the public at large is made to live with. Once in place, a certain governance trajectory is difficult to change due to the tendency of regulatory institutions to enshrine a certain power distribution that becomes more difficult to overcome as time passes. Hence, it is paramount to include citizens, foster state expertise, and face critics' concerns in the earlier stages of technology regulation in order to avoid a bias in regulation in favor of industry interests. These are important lessons at a time when new technologies promise to transform the world we live in, and safety governance choices now will shape future politics in each of these areas.

Notes

INTRODUCTION

1. The stress tests conducted by the European Nuclear Safety Regulators Group as a "reassessment of the safety margins" (ENSREG 2011) deemed Europe's nuclear power plants safe. Safety margins refer to safety measures that allow a plant operator to cool the reactor core and prevent the emission of radioactive particles after existing safety systems have failed to prevent an accident. This narrow definition of a safety stress test, excluding a review of preventative measures and other factors, such as human error, elicited criticism from civil society organizations, such as Greenpeace (Wenisch and Becker 2012), as well as the Greens in the European Parliament (Renneberg 2011).

2. The term "nuclear village" was coined by Tetsunari Iida in the late 1990s. He left the nuclear industry to become an avocate of renewable energy development and is executive director of Japan's Institute for Sustainable Energy Policies (ISEP).

3. The Liberal Democratic Party has been a staunch supporter of nuclear power development virtually since its formation in 1955. It has always been Japan's largest party and has dominated government most years since then. The exceptions are the years 1993–1994 and 2009–2012, when the LDP was in the opposition.

4. Many attribute this aphorism to the British philosopher Francis Bacon. Bacon's "nam et ipsa scientia potestas est" (*De Haeresibus* 1597) translates into "for knowledge itself is power."

1. THE "NUCLEAR VILLAGE"

1. Iida left the nuclear industry to become an advocate of renewable energy development. He is executive director of the Institute for Sustainable Energy Policies in Japan.

2. Each electric utility is responsible for providing electricity to a certain area and is organized as a regional monopoly. The nine electric utilities are Tokyo Electric Power Company (TEPCO), Kansai Electric Power Company (KEPCO), Tohoku Electric Power Company, Hokkaido Electric Power Company, Hokuriku Electric Power Company, Chugoku Electric Power Company, Kyushu Electric Power Company, Shikoku Electric Power Company, and Chubu Electric Power Company. When the United States returned Okinawa to Japan in 1972, the Okinawa Electric Power Company became the tenth electric utility in Japan. Okinawa's electric utility refrained from developing nuclear power. Furthermore, there is J-Power, which operates the transmission lines connecting the four main islands. J-Power is not operating any nuclear power plants, but as of 2022 still has one under construction. Not an electric utility in itself, Japan Atomic Power Co. (JAPC) is a company founded in order to develop commercial nuclear power in Japan and is jointly owned by some of Japan's electric utilities.

3. The *Yomiuri Shimbun* is the largest of five nationwide Japanese newspapers. It has a combined morning and evening circulation of about 13 million copies.

4. The Japanese slogan "*Genshiryoku-akarui-mirai*" decorated a street of Futaba, a town close to the damaged Fukushima Daiichi nuclear power plant. It was widely depicted in the media after the accident. Since the town had to be evacuated, the sign became a cynical reminder of what nuclear power stood for before it became a ghost town.

The sign was removed in 2016, but the municipality intended to preserve it for an accident remembrance park.

5. It was quasidomestic, because uranium needed to be imported.

6. The *Asahi Shimbun* is one of five nationwide daily newspapers in Japan. The morning and evening edition together reach a circulation of close to 11 million copies. Its political orientation is center-left.

7. This number counts lawsuits against commercial nuclear power plants. Appeals before higher courts are not counted separately. The number excludes lawsuits related to the fast breeder reactor Monju, enriching uranium, using mixed oxide fuels, the JCO accident, workers' exposure to radiation, and radioactive waste storage.

8. One case, the last pre-3.11 lawsuit, filed in 2010 against the construction of the Oma NPP, was ongoing at the time of writing in 2019.

9. This is a translation of a well-known verse of German humorist Christian Morgenstern: "Weil . . . nicht sein kann, was nicht sein darf" (1964).

10. Sometimes also translated as "Basic Energy Plan."

2. 3.11 AS AN OPPORTUNITY FOR CHANGE

1. The term "new energy sources" is used in Japanese energy policy documents to denote both renewable energy and nuclear power.

2. The round of experts consisted of (in alphabetical order) Iida Tetsunari (Executive Director of the Institute for Sustainable Energy Policies), Iizuka Yoshinori (Professor at the Graduate School of Engineering, The University of Tokyo), Ikawa Yōjirō (Editorial Writer at The Yomiuri Shimbun), Kawakatsu Heita (Governor of Shizuoka Prefecture), Kitamura Masaharu (Professor Emeritus from Tōhoku University), Sekimura Naoto (Vice Dean and Professor at Graduate School of Engineering, The University of Tokyo), Sumita Hiroko (Attorney-at-law), Suzuki Motoyuki (Professor Emeritus from The University of Tokyo and Chairman of the Central Environmental Council), Sudō Yuki (Director of the Research Institute of Social Safety), and Takahashi Shigeru (Director of the School of International and Public Policy (IPP), Hitotsubashi University).

3. Iida is the head of the Institute for Sustainable Energy Policies. After leaving the nuclear industry in the 1990s, he coined the term "nuclear village."

4. A backfit system refers to retrofitting existing reactors in line with newest scientific findings and updates in safety measures.

5. For a good overview of these reports see Lukner and Sakaki (2013).

6. With short interruptions in 1993–94 and 2009–2012, the LDP governed Japan alone or in coalition governments.

7. Deliberative polling follows a method developed at Stanford University where participants are polled to begin with, then debate the issue in question and eventually are polled again.

3. THE NUCLEAR REGULATION AUTHORITY (NRA)

1. Fuketa Toyoshi is a nuclear engineer, who had previously worked at the JAEA (former JAERI) from 1987.

2. Nakamura Kayoko, a specialist for nuclear medicine and radioisotopes, had previously worked at the Japanese Society of Nuclear Medicine from 1994 and joined the Japan Radioisotopes Organization in 2010.

3. Ōshima Kenzō held a degree in law and international relations. He had previously held a position as the UN Undersecretary-General for Humanitarian Affairs (2001–03) and was a member of the Nuclear Accident Independent Investigation Committee in 2011–12.

4. Shimazaki Kunihiko, a seismologist, was a professor at the Tokyo University Seismology Institute from 1989 and the chairman of the Coordinating Committee for Earthquake Prediction (CCEP) at the Ministry of Land, Infrastructure and Transport from 2009.

5. Tanaka Shun'ichi is a nuclear engineer, who had previously worked at the JAEA (formerly JAERI) from 1967.

6. In 2005, JAERI was merged with the Japan Nuclear Cycle Development Institute to form the Japan Atomic Energy Agency (JAEA).

7. The safety agency's YouTube channel, "NRAJapan," became a window into NRA board meetings and other meetings, which were usually streamed live with recordings available afterwards.

8. Study teams consist of at least one board member, along with NRA Secretariat officials. Sometimes they include external experts or members of incorporated administrative agencies or semiprivate research institutes, such as JNES (until its integration into the NRA) and JAEA.

9. Online, the NRA provides PDF documents with self-reported data by each external expert regarding their affiliation and the amount of money received from companies with ties to the nuclear industry and so forth. While the names in the documents have been blackened out, the link to the respective document contains the name of the respective external expert, meaning it can be easily attributed to the person.

10. Expert meetings include many external experts and only a few NRA officials or board members. Invited experts come from a variety of related research institutes and university departments and have to meet the neutrality criteria above. Such expert meetings convene whenever the NRA seeks external input on a specific issue.

11. For these, a short summary with a list of participants, the topic under discussion, and a few bullet points about the discussion became available on the website.

12. The number does not include local councils and committees that NRA Secretariat officials attend.

13. Calculated by the author based on data obtained during an interview with an NRA official on August 18, 2015.

14. Concretely, it covers ANRE's General Policy Division (excluding the budget and administrative affairs office), ANRE's Electricity and Gas Industry Department (excluding the Gas Market Division), executive positions within ANRE, MEXT's Research and Development Bureau Research and Development Policy Division, MEXT's Environment and Energy Division (limited to the director of the division and the Office for Fusion Energy), MEXT's Atomic Energy Division, AEC's Bureau of the Atomic Energy Commission.

15. Tanaka himself stressed that no such thing as absolute safety exists with a high-risk technology such as nuclear power. Rather, one can only lower the risks by improving nuclear safety.

4. POST-3.11 NUCLEAR SAFETY STANDARDS

1. Study teams included at least one NRA board member, external experts, NRA Secretariat officials, and sometimes members of incorporated administrative agencies or semiprivate research institutes, such as JNES (before its integration into the NRA) and the JAEA. To develop new safety requirements and a safety review system, the NRA established the following study teams: Study Team on New Regulatory Requirements for Light Water Power Reactors (10/2012–06/2013), Study Team on New Regulatory Requirements for Light Water Nuclear Power Plants (Earthquakes and Tsunamis) (11/2012–06/2013), Study Team on New Regulatory Requirements for Nuclear Fuel Facilities

(04/2013–10/2013), Study Team on Radiation Emergency Medicine (11/2012–08/2015), Study Team on Nuclear Emergency Preparedness Measures (11/2012–08/2015), Study Team on Emergency Monitoring (12/2012–03/2013), Study Team on a Review System for New Safety Regulations for Light Water Nuclear Power Plants (11/2012–10/2013)

2. Regulatory requirements for nuclear fuel cycle facilities were developed separately with individual requirements for each facility due to vast differences in construction design.

3. For a good overview of nuclear security and safeguarding efforts and how they are linked to nuclear safety, see Wyss (2016) and Naito (2016).

4. In line with different requirements for commercial nuclear power plants and nuclear fuel cycle facilities, there are Review Meetings on Conformity to the New Regulatory Requirements for Nuclear Power Plants and Review Meetings on Conformity to the New Regulatory Requirements for Nuclear Fuel Facilities.

5. The Japanese fiscal year runs from April 1 of the stated year until March 31 of the following year.

5. THE FISSURED "NUCLEAR VILLAGE"

1. Ishiwatari Akira received a PhD in geology from the University of Tokyo and, in 2008, became a professor at the Tōhoku University. Also, he served as president of the Geological Society of Japan from 2012 until 2014.

2. Tanaka Satoru received a PhD in engineering from the University of Tokyo. He is an expert on nuclear fuel cycles and radioactive waste. He became a professor in the Department of Nuclear Engineering and Management, School of Engineering at the University of Tokyo in 2008. He left this post to become an NRA commissioner in 2014. In addition, he became president of the Atomic Energy Society of Japan (AESJ) in 2011. In this capacity he created the Fukushima Special Project and was chairman of AESJ's Investigation Committee on the Fukushima Daiichi Nuclear Accident.

3. On the Japanese website, the designated header was "accidents and safety" (*Jiko to anzen*). On the English website, there were two designated headers, "nuclear regulation" and "safety."

4. The numbers in this section refer to lawsuits against commercial nuclear power plants. Appeals before higher courts are not counted separately. Numbers exclude lawsuits related to the fast breeder reactor Monju, enriching uranium, using mixed oxide fuels, the JCO accident, workers' exposure to radiation, and radioactive waste storage.

5. The wave of lawsuits following 3.11 was not limited to cases against commercial nuclear power plants. For an overview of accident-related lawsuits, such as liability cases for the loss of homes after evacuation, please see Togni (2022).

6. The lawsuit against the Ōi plant in adjoining Kyoto Prefecture was filed before the new safety standards went into effect, which makes it an instance of widened antinuclear protests after 3.11, but not an example of protests galvanized by the NRA.

References

Abe, Shinzo. 2013. "Press Conference by Prime Minister Shinzo Abe: Friday, January 4, 2013." Accessed October 6, 2022. http://japan.kantei.go.jp/96_abe/statement/201301 /04kaiken_e.html.

Ahn, Joonhong, Cathryn Carson, Mikael Jensen, Kohta Juraku, Shinya Nagasaki, and Satoru Tanaka, eds. 2015. *Reflections on the Fukushima Daiichi Nuclear Accident: Toward Social-Scientific Literacy and Engineering Resilience.* Cham: Springer Open.

Al-Badri, Dominic, and Gijs Berends, eds. 2013. *After the Great East Japan Earthquake: Political and Policy Change in Post-Fukushima Japan.* Asia Insights 5. Copenhagen: NIAS Press.

Aldrich, Daniel P. 2010. *Site Fights: Divisive Facilities and Civil Society in Japan and the West.* Ithaca, NY: Cornell University Press.

Aldrich, Daniel P. 2013. "Revisiting the Limits of Flexible and Adaptive Institutions: The Japanese Government's Role in Nuclear Power Plant Siting over the Post-War Period." In Jeff Kingston, ed., *Critical Readings on Japan,* 79–91. New York: Routledge.

Aldrich, Daniel P., and Timothy Fraser. 2017. "All Politics Is Local: Judicial and Electoral Institutions' Role in Japan's Nuclear Restarts." *Pacific Affairs* 90 (3): 433–57. https://doi.org/10.5509/2017903433.

Amano, Kensuke. 2015. *Genshiryoku-Kisei-Iinkai No Kodoku: Genpatsu Saikadō No Shinsō* [The solitude of the NRA: The truth behind nuclear restarts]. Enerugī fōramu shinsho 31. Tokyo: Enerugīfōramu.

ANRE. 2015. "Genshiryoku-Hatsuden Ni Okeru Ronten" [Nuclear power issues].

Arima, Tetsuo. 2008. *Genpatsu, Shōriki, CIA: Kimitsu Bunsho De Yomu Shōwa Rimenshi* [Nuclear power plants, Shōriki and the CIA: Reading the inside stories of Shōwa through secret documents], 249. Tōkyō: Shinchōsha.

Arnhold, Valerie. 2021. "Normalisation of Nuclear Accidents After the Cold War." *Cold War History* 21 (3): 261–81. https://doi.org/10.1080/14682745.2020.1806239.

Asahi Shimbun. 2011a. "Genpatsu Shin-Soshiki, Tsunahiki: Tantōshō Shian" [A new nuclear power agency: A tug-of-war between different proposals]. August 6, morning edition.

Asahi Shimbun. 2011b. "Keizaishō" Haijo O Yūsen: Genshiryoku Kisei, Kankyōshō, Jinzai Kakuho Kagi" [Prioritizing the separation from "METI": Nuclear regulation, the Ministry of the Environment and the key to securing skilled personnel]. August 12, morning edition.

Asahi Shimbun. 2012a. "Genpatsutomedia: Genshiryokumurawomegutte" [Nuclear power plants and the media: On the nuclear village]. March 2, morning edition.

Asahi Shimbun. 2012b. "Genshiryoku Kisei-Chō, Mienu Hōan Seiritsu: Yongatsu Tsuitachi Hossoku, Kishingō" [Nuclear regulation agency legislation approval nowhere in sight: Yellow light for the April 1st inauguration]. February 24, morning edition.

Asahi Shimbun. 2012c. "Kisei-Chō, Tsuitachi Hossoku Dannen, Seiken, Genpatsu Kanshi, Tōmen Wa Hoanin" [Aborted April 1 inauguration of the regulation agency: Government entrusts supervision of nuclear plants to NISA for time being]. March 22, morning edition.

Asahi Shimbun. 2012d. "Kisei-Chō, Yōyaku Ronsen, Seiken, Hōan Shūsei Ni Mae-muki" [Debate on nuclear regulation agency finally underway: Government optimistic about legislation reform]. May 30, morning edition.

Asahi Shimbun. 2012e. "Kisei-Chō Dokuritsu, Chiji "Nozomashii": 3 Tō Shūsei Kyōgi Gōi De" [Governor [says] regulation agency's independence is "desirable": Three parties hold meeting and reach agreement on revision]. June 7, morning edition.

Asahi Shimbun. 2012f. "Genpatsu, Shushō No Shijiken Gentei: 3 Tō Gōi, Kisei-I Hōan, Setsuritsu E" [Prime minister's authority over power plants restricted: Three parties reach agreement on bill for establishment of regulation authority]. June 12, morning edition.

Asahi Shimbun. 2012g. "Genshiryoku Mura Wa Taishōgai": Kisei-I Jinji Meguri Kankyōshō" [Environment minister on staffing the regulation commission: "Nuclear village will be excluded"]. June 19, morning edition.

Asahi Shimbun. 2012h. "Genshiryoku Kisei-I "Mura Haijo" No Jinsen Shishin, "3 Nen Nai Ni Genshiryoku Kanren Yakuin" Wa Jogai Nado" [NRA instructed to "exclude the village" when selecting members: "People involved with nuclear industry in the past three years" also to be excluded]. July 3, morning edition.

Asahi Shimbun. 2012i. "Datsu-Genshiryoku Mura" De Jinsen Kisei-I Toppu Ni Tanaka Shunichi-Shi Kiyō E Kokkai Teiji Wa Sakiokuri" [Member selection based on (the principle of) "abandoning the nuclear village." Diet proposal of appointing Mr. Tanaka Shunichi as head of NRA postponed]. July 21, morning edition.

Asahi Shimbun. 2012j. "Genshiryoku Kisei-I Jinji No Henkō Yōkyū, Minshu PT" [DPJ project team demands changes to NRA staff]. August 2, evening edition.

Asahi Shimbun. 2012k. "Genshiryoku-Kisei'ira, Shushōninmei E: Kokkai No Dōi Ezu" [On the path to NRA commission members appointed by the prime minster: No Diet approval]. September 5, morning edition.

Asahi Shimbun. 2014a. "Pro-Nuclear Expert Who Raised Flag on Quake Risk." May 28, online edition.

Asahi Shimbun. 2014b. "Kiseii No Dokuritsu Ayabumu Koe" [Concerns voiced about regulators' independence]. June 11, morning edition.

Asahi Shimbun. 2014c. "New Nuclear Watchdog Commissioner Received Additional Industry Payment." July 5, online edition.

Asahi Shimbun. 2015. "Government Electricity Targets to Support Nuclear Industry." August 13, online edition.

Asahi Shimbun. 2016a. "Takahamagenpatsusanyongoki, Untensashitome: Ootsuchisaiga-karishobun" [Ōtsu district court issues temporary injunction, prohibiting operation of Units 3 and 4 at Takahama nuclear power plant]. March 9, online edition.

Asahi Shimbun. 2016b. "Asahi Shimbunsha Yoronchōsa" [Asahi Newspaper Company public opinion poll]. February 16, morning edition.

Asahi Shimbun. 2017. "District court rejects request to temporarily halt Ikata nuke plant." March 30, online edition.

Asahi Shimbun. 2018. "Saikadoumitomenuhanketsu: Saibanchounoshinnen 'Kikaseta-nowaaikokushin'" [Chief judge beliefs decision to halt nuclear reactor restart was "act of patriotism"]. August 4, online edition. Accessed October 6, 2022. https://www.asahi.com/articles/ASL824FPWL82UPQJ00C.html.

Asahi Shimbun. 2019a. "Genpatsuanzentaisakuhi, 5chouenchouni: Seifuno "Saiyasu" Hyoukayuragu" [Costs for nuclear safety measures over 5 trillion yen: Questioning government's appraisal as "cheapest"]. August 12, online edition. Accessed October 6, 2022. https://www.asahi.com/articles/ASM7R6KNCM7RULBJ00S.html.

Asahi Shimbun. 2019b. "Yoronchousa-Shitsumon to Kotae, Fukushimaken, 2 Gatsu 23, 24 Nichi Jitsushi" [Public opinion poll questions and answers, Fukushima Prefecture, conducted on February 23 and 24]. February 27, online edition. Accessed October 6, 2022. https://www.asahi.com/articles/ASM2T4TW2M2TUZPS007.html.

Asahi Shimbun. 2022. "Genpatsu Saikai, Hantai Wa Hansūware 47%, Asahi Yoronchōsa" [Opposition to nuclear reactor restarts down to 47% According to Asahi poll]. February 22, online edition. Accessed October 6, 2022. https://www.asahi.com/articles/ASQ2P5V3GQ2PUZPS006.html.

Avenell, Simon. 2012. "From Fearsome Pollution to Fukushima: Environmental Activism and the Nuclear Blind Spot in Contemporary Japan." *Environmental History* 17 (2): 244–76. https://doi.org/10.1093/envhis/emr154.

Balleisen, Edward J., Lori S. Bennear, Kimberly D. Krawiec, and Jonathan B. Wiener, eds. 2017. *Policy Shock: Recalibrating Risk and Regulation After Oil Spills, Nuclear Accidents, and Financial Crises.* Cambridge: Cambridge University Press.

Ban, Hideyuki. 2015. Interview by F. Koppenborg. Tokyo, July 29.

Ban, Hideyuki. 2018. Interview by F. Koppenborg. Tokyo, August 22.

Baumgartner, Frank R., and Bryan D. Jones. 1993. *Agendas and Instability in American Politics.* American Politics and Political Economy Series. Chicago: University of Chicago Press.

BBC. 2011. "Japan Earthquake: Meltdown Alert at Fukushima Reactor." https://www.bbc.com/news/world-asia-pacific-12733393.

Beck, Ulrich. 1992. *Risk Society: Towards a New Modernity.* Theory, Culture & Society. London: Sage Publications.

Béland, Daniel, Martin B. Carstensen, and Leonard Seabrooke. 2016. "Ideas, Political Power and Public Policy." *Journal of European Public Policy* 23 (3): 315–17. https://doi.org/10.1080/13501763.2015.1122163.

Benford, Robert D., and David A. Snow. 2000. "Framing Processes and Social Movements: An Overview and Assessment." *Annual Review of Sociology* 26: 611–39. https://doi.org/10.1146/annurev.soc.26.1.611.

Bengodan. 2020. "About Assembly of Japan's Lawyers for a Nuclear Phase Out." Accessed July 20, 2020. http://www.datsugenpatsu.org/bengodan/about/.

Berlin, Isaiah. 1974. "Historical Inevitability." In *The Philosophy of History*, edited by Patrick Gardiner, 161–86. Oxford readings in philosophy. Oxford: Oxford University Press.

Bianculli, Andrea C., Jacint Jordana, and Ana G. Juanatey. 2017. "International Networks as Drivers of Agency Independence: The Case of the Spanish Nuclear Safety Council." *Administration & Society* 49 (9): 1246–71. https://doi.org/10.1177/0095399715581034.

Birkland, Thomas A. 2007. *Lessons of Disaster: Policy Change After Catastrophic Events.* 1st ed. American Governance and Public Policy Series. Washington, DC: Georgetown University Press.

Blandford, Edward David, and Scott Douglas Sagan, eds. 2016. *Learning from a Disaster: Improving Nuclear Safety and Security After Fukushima.* Stanford, CA: Stanford University Press.

Blatter, Joachim, and Markus Haverland. 2012. *Designing Case Studies: Explanatory Approaches in Small-N Research.* ECPR Research Methods Series. London: Palgrave Macmillan.

Blyth, Mark. 2002. *Great Transformations: Economic Ideas and Institutional Change in the Twentieth Century.* New York: Cambridge University Press.

Brown, Alexander James. 2018. *Anti-Nuclear Protest in Post-Fukushima Tokyo: Power Struggles.* 1st ed. Routledge/Asian Studies Association of Australia (ASAA) East Asian Series. London: Taylor and Francis.

Cabinet Secretariat. 2011a. "'Genshiryoku Anzen Kisei Soshiki Tō Kaikaku Junbishitsu' No Secchi" [Establishment of "Preparation Office for Nuclear Safety Regulation and Reform"]. Accessed October 23, 2016. http://www.cas.go.jp/jp/genpatsujiko /info/news_110826.html.

Cabinet Secretariat. 2011b. "Recommendations from Advisory Committee for Prevention of Nuclear Accident." December 13. Accessed October 6, 2022. http://www .cas.go.jp/jp/genpatsujiko/info/news_110930.html.

Cabinet Secretariat. 2011c. "Review on an Organization in Charge of Nuclear Safety Regulation." Accessed October 23, 2016. http://www.cas.go.jp/jp/genpatsujiko/info /shian_en_110805.html.

Cabinet Secretariat. 2011d. "Review on an Organization in Charge of Nuclear Safety Regulation: Understanding by Relevant Ministers." Accessed October 6, 2022. http:// www.cas.go.jp/jp/genpatsujiko/index.html.

Cabinet Secretariat. 2012a. "Dai Ikkai Genshiryoku Kisei I'inkai He No Dokuritsugyōseihōjin Genshiryokuanzenkibankikō No Sōgō Ni Kansuru Fukudaijin Nado Kaiken Gijiyōshi" [The gist of first press conference by state minister on integrating JNES into the Nuclear Regulation Authority]. Accessed October 6, 2022. https://www.cas .go.jp/jp/genpatsujiko/jnes1/100.html.

Cabinet Secretariat. 2012b. "Genshiroyku No Anzen No Kakuho Ni Kan Suru Sohiki Oyobi Seido O Kaikaku Suru Tame No Kankyōshō Secchihō No Ichibu O Kaisei Suru Hōritsu-an" [Draft bill for revising parts of Ministry of Environment's Establishment Law in order to reform institutions and systems related to ensuring nuclear safety]. Accessed October 6, 2022. http://www.cas.go.jp/jp/genpatsujiko /info/siryo0131/tabane_yoko.pdf.

Cabinet Secretariat. 2012c. "Reform of Japan's Nuclear Safety Regulation: January 2012." Accessed October 6, 2022. https://www.cas.go.jp/jp/genpatsujiko/info/kokusaiws /siryo/reform_of_regulation.pdf .

Cabinet Secretariat. 2012d. "Report of the International Workshop on Nuclear Safety Regulation: 18th January 2012, Tokyo/Japan." Accessed October 6, 2022. https:// www.cas.go.jp/jp/genpatsujiko/info/report/report_en.pdf.

Cabinet Secretariat. 2016. "Press Conference by Prime Minister Shinzo Abe on the Upcoming Fifth Anniversary of the Great East Japan Earthquake." March 10. Accessed October 6, 2022. http://japan.kantei.go.jp/97_abe/statement/201603/1216516_11003 .html.

Calder, Kent E. 1988. *Crisis and Compensation: Public Policy and Political Stability in Japan, 1949–1986*. Princeton, NJ: Princeton University Press.

Capoccia, Giovanni. 2016a. "Critical Junctures." In The *Oxford Handbook of Historical Institutionalism*, edited by Orfeo Fioretos, Tulia G. Falleti, and Adam Sheingate, 89–106. Oxford Handbooks. Oxford: Oxford University Press.

Capoccia, Giovanni. 2016b. "When Do Institutions "Bite"? Historical Institutionalism and the Politics of Institutional Change." *Comparative Political Studies* 49 (8): 1095–1127. https://doi.org/10.1177/0010414015626449.

Capoccia, Giovanni, and R. Daniel Kelemen. 2007. "The Study of Critical Junctures: Theory, Narrative, and Counterfactuals in Historical Institutionalism." *World Politics* 59 (3): 341–69. https://doi.org/10.1017/s0043887100020852.

Carpenter, Daniel. 2010. *Reputation and Power: Organizational Image and Pharmaceutical Regulation at the FDA*. Princeton Studies in American Politics, vol. 137. Princeton, NJ: Princeton University Press.

Carpenter, Daniel P. 2001. *The Forging of Bureaucratic Autonomy: Reputations, Networks, and Policy Innovation in Executive Agencies, 1862–1928*. Princeton, NJ: Princeton University Press.

Carr, Edward Hallett. 1961. *What Is History?* London: Penguin Books.

Caruso, Gustavo. 2012. "Presentation IAEA: IAEA Safety Standards and Experiences." Accessed October 6, 2022. https://www.cas.go.jp/jp/genpatsujiko/info/kokusaiws/siryo/iaea.pdf.

Chiavacci, David, and Julia Obinger. 2018. "Towards a New Protest Cycle in Contemporary Japan? Resurgence of Social Movements and Confrontational Political Activism in Historical Perspective." In *Social Movements and Political Activism in Contemporary Japan: Re-Emerging from Invisibility*, edited by David Chiavacci and Julia Obinger, 1–23. The Mobilization Series on Social Movements, Protest, and Culture. Abingdon and New York: Routledge.

CNIC. 2014. "News Watch 160 May/June 2014 Nuke Info Tokyo No. 160: Hakodate City Files Suit to Halt Construction of Ohma Nuclear Power Station." Accessed October 23, 2016. http://www.cnic.jp/english/?p=2926#3.

CNIC. 2015. "Extending the Lifespan of Takahama Units 1 and 2: Sloppy RPV Surveillance Method." Accessed October 23, 2016. http://www.cnic.jp/english/?p=3117.

CNIC. 2020, personal communications regarding data on lawsuits against nuclear power plants in Japan, April 8.

Cohen, Linda, Mathew D. McCubbins, and Frances McCall Rosenbluth. 1995. "The Politics of Nuclear Power in Japan and the United States." In *Structure and Policy in Japan and the United States*, edited by Peter F. Cowhey and Mathew D. McCubbins, 177–202. Cambridge and New York: Cambridge University Press.

Collier, David. 2011. "Understanding Process Tracing." *APSC (Political Science & Politics)* 44 (4): 823–30. https://doi.org/10.1017/S1049096511001429.

Collier, David, and Gerardo L. Munck. 2017. "Building Blocks and Methodological Challenges: A Framework for Study in Critical Junctures." *Qualitative and Multi-Method Research* 15 (1): 2–9.

Cotton, Matthew. 2014. "Structure, Agency and Post-Fukushima Nuclear Policy: An Alliance-Context-Actantiality Model of Political Change." *Journal of Risk Research* 18 (3): 1–16. https://doi.org/10.1080/13669877.2014.919512.

Culpepper, Pepper D. 2010. *Quiet Politics and Business Power: Corporate Control in Europe and Japan*. Cambridge Studies In Comparative Politics. Cambridge: Cambridge University Press.

Denjiren. 2011a. "Denjirenkaichō: Teireikaikenyōshi" [Chairman of Denjiren: Regular press conference (summary)]. July 15. Accessed October 6, 2022. http://www.fepc.or.jp/about_us/pr/kaiken/__icsFiles/afieldfile/2011/07/15/kaiken20110715.pdf.

Denjiren. 2011b. "Denjirenkaichō: Teireikaikenyōshi" [Chairman of Denjiren: Regular press conference (summary)]. April 15. Accessed October 6, 2022. http://www.fepc.or.jp/about_us/pr/kaiken/__icsFiles/afieldfile/2011/04/15/kaiken_0415.pdf.

Denjiren. 2012. "FEPC's Stance on the Decision of 'Innovative Strategy for Energy and Environment': September 14, 2012." Accessed October 6, 2022. https://www.fepc.or.jp/english/news/conference/__icsFiles/afieldfile/2012/09/26/kaiken_E_120914_f.pdf.

Denjiren. 2013. "Summary of Press Conference Comments Made by Makoto Yagi, FEPC Chairman: On November 15, 2013." Accessed October 6, 2022. https://www.fepc.or.jp/english/news/conference/__icsFiles/afieldfile/2013/11/20/kaiken_e_20131115.pdf.

Denjiren. 2014a. "Denjirenkaichō: Teireikaikenyōshi" [Chairman of Denjiren: Regular press conference (summary)]. April 18. Accessed October 6, 2022. http://www.fepc.or.jp/about_us/pr/kaiken/__icsFiles/afieldfile/2014/04/18/kaiken_20140418.pdf.

Denjiren. 2014b. "Denjirenkaichō: Teireikaikenyōshi" [Chairman of Denjiren: Regular press conference summary]. May 23. Accessed October 23, 2016. http://www.fepc.or.jp/about_us/pr/kaiken/__icsFiles/afieldfile/2014/05/23/kaiken_20140523.pdf.

Denjiren. 2014c. "Denjirenkaichō: Teireikaikenyōshi" [Chairman of Denjiren: Regular press conference (summary)]. November 14. Accessed October 6, 2022. http://www .fepc.or.jp/about_us/pr/kaiken/__icsFiles/afieldfile/2014/11/14/kaiken_20141114 .pdf.

Denjiren. 2015a. "Summary of Press Conference Comments Made by Makoto Yagi, FEPC Chairman: On February 20, 2015." Accessed October 6, 2022. https://www.fepc.or .jp/english/news/conference/2014.html.

Denjiren. 2015b. "Summary of Press Conference Comments Made by Makoto Yagi, FEPC Chairman: On March 20, 2015." Accessed October 6, 2022. https://www.fepc.or .jp/english/news/conference/2014.html.

Denjiren. 2016. "Denjirenkaichō: Teireikaikenyōshi" [Chairman of Denjiren: Regular press conference (summary)]. July 15. Accessed October 6, 2022. http://www.fepc .or.jp/about_us/pr/kaiken/__icsFiles/afieldfile/2016/07/15/kaiken_20160715.pdf.

Denjiren. 2017. "Summary of Press Conference Comments Made by Satoru Katsuno, FEPC Chairman: On September 15, 2017." Accessed October 6, 2022. https://www .fepc.or.jp/english/news/conference/2017.html.

Denjiren. 2018a. "Summary of Press Conference Comments Made by Satoru Katsuno, FEPC Chairman: On July 20, 2018." Accessed October 6, 2022. https://www.fepc .or.jp/english/news/conference/2018.html.

Denjiren. 2018b. "Summary of Press Conference Comments Made by Satoru Katsuno, FEPC Chairman: On May 18, 2018." Accessed October 6, 2022. https://www.fepc .or.jp/english/news/conference/2018.html.

Denjiren. 2019. "Summary of Press Conference Comments Made by Shigeki Iwane, FEPC Chairman: On September 20, 2019." Accessed October 6, 2022. https:// www.fepc.or.jp/english/news/conference/2019.html.

DeWit, Andrew. 2014. "Japan's Renewable Power Prospects." In Kingston 2014, 120–34.

DiMaggio, Paul, and Walter Powell. 1983. "The Iron Cage Revisited: Institutional Isomorphism and Collective Rationality in Organizational Fields." *American Sociological Review* 48 (2): 147–60. http://www.jstor.org/stable/2095101?origin=JSTOR -pdf.

DPJ Government Adviser. 2015. Interview by F. Koppenborg. June 15, Tokyo.

Durant, Robert F. 2004. "The Precautionary Principle." In *Environmental Governance Reconsidered: Challenges, Choices, and Opportunities*, edited by Robert F. Durant, Daniel J. Fiorino, and Rosemary O'Leary, 105–44. American and Comparative Environmental Policy. Cambridge, MA: MIT Press.

Eberlein, Burkhard, and Abraham L. Newman. 2008. "Escaping the International Governance Dilemma? Incorporated Transgovernmental Networks in the European Union." *Governance* 21 (1): 25–52. https://doi.org/10.1111/j.1468-0491 .2007.00384.x.

Economic News. 2017. "Keidanren Ga Genpatsu Ripre-Su/shinzōsetu Wo Unagasu" [Keidanren calling for repowering old reactors and building new ones]. June 6. http://economic.jp/?p=74522.

Edahiro, Junko. 2012. "Update on the Discussion in Japan on Energy and Environment Policy Options to 2030." Accessed October 6, 2022. https://www.japanfs.org/en /news/archives/news_id032229.html.

Elliott, David. 2013. *Fukushima: Impacts and Implications*. Palgrave Pivot. Basingstoke, UK: Palgrave Macmillan.

Endo, Masahisa, Robert Pekkanen, and Steven R. Reed. 2013. "The LDP's Path Back to Power." In *Japan Decides 2012: The Japanese General Election*, edited by Robert Pekkanen, Steven R. Reed, and Ethan Scheiner, 49–64. Basingstoke, UK: Palgrave Macmillan.

ENSREG. 2011. "Declaration of ENSREG: EU 'Stress Tests' Specifications." Accessed October 6, 2022. http://www.ensreg.eu/node/286.

Environmental Policy Expert. 2014. Interview by F. Koppenborg. September 18., Tokyo.

Ethik-Kommission Sichere Energieversorgung. 2011. *Deutschlands Energiewende: Deutschlands Energiewende–EinGemeinschaftswerk Für Die Zukunft*. Accessed October 6, 2022. https://www.nachhaltigkeitsrat.de/wp-content/uploads/migration /documents/2011-05-30-abschlussbericht-ethikkommission_property_publication File.pdf.

EU Commission. 2011. "Communication from the Commission to the Council and the European Parliament on the Interim Report on the Comprehensive Risk and Safety Assessments 'Stress Tests' of Nuclear Power Plants in the European Union: COM/2011/0784." Accessed October 6, 2022. https://eur-lex.europa.eu/legal -content/EN/TXT/HTML/?uri=CELEX:52011DC0784&from=GA.

Fearon, James D. 1991. "Counterfactuals and Hypothesis Testing in Political Science." *World Politics* 43 (2): 169–95. https://doi.org/10.2307/2010470.

Ferguson, Charles D., and Mark Jansson. 2013. *Regulating Japanese Nuclear Power in the Wake of the Fukushima Daiichi Accident*. Accessed October 6, 2022. http://fas.org /wp-content/uploads/2013/05/Regulating_Japanese_Nuclear_13May131.pdf.

Fioretos, Orfeo. 2011. "Historical Institutionalism in International Relations." *International Organization* 65 (02): 367–99. https://doi.org/10.1017/S0020818311000002.

Former Administrative Vice Minister. 2015. Interview by F. Koppenborg. August 3. Tokyo.

Former AEC Official. 2015. Interview by F. Koppenborg. July 31. Tokyo.

Former AEC Official. 2016. Interview by F. Koppenborg. January 13. Berlin.

Former Board Member. 2019. Interview by F. Koppenborg. April 11. Iitatemura.

Former NSC Member. 2016. Interview by F. Koppenborg. April 1. Tokyo.

Freeman, Laurie Anne. 2000. *Closing the Shop: Information Cartels and Japan's Mass Media*. Princeton, NJ: Princeton University Press.

Gilardi, Fabrizio. 2008. *Delegation in the Regulatory State // Delegation in the Regulatory State: Independent Regulatory Agencies in Western Europe // Independent Regulatory Agencies in Western Europe*. Cheltenham, UK: Edward Elgar Publishing.

Gilardi, Fabrizio, and Martino Maggetti. 2011. "The Independence of Regulatory Agencies." In Levi-Faur 2011, 201–14.

Greenpeace Japan. 2016a. "Japan's Nuclear Regulator Rubberstamps Restarts Before Finishing Safety Reviews." Accessed October 23, 2016. http://www.greenpeace.org /japan/ja/news/press/2016/pr201604201/.

Greenpeace Japan. 2016b. "Press Release English." Accessed October 23, 2016. http://www .greenpeace.org/japan/ja/news/PR-english/.

Greenpeace Japan. 2016c. "Puresu Riri-Su: Press Release." Accessed October 23, 2016. http://www.greenpeace.org/japan/ja/news/press/.

Groenleer, M. L. P. 2012. "Linking up Levels of Governance: Agencies of the European Union and International Institutions." In *The Influence of International Institutions on the EU: When Multilateralism Hits Brussels*, edited by Oriol Costa and K. E. Jorgensen, 135–54. Palgrave Studies in European Union Politics. Basingstoke, UK: Palgrave Macmillan.

Guardian. 2011a. "News Blog: Japan Tsunami and Nuclear Alert." March 14. Accessed October 6, 2022. https://www.theguardian.com/world/2011/mar/14/japan-tsunami -nuclear-alert-live-coverage.

Guardian. 2011b. "Britain Joins Countries Urging Their Citizens to Leave Tokyo." March 16, online edition. Accessed October 6, 2022. https://www.theguardian .com/world/2011/mar/16/britain-urging-citizens-leave-tokyo.

Guidi, Mattia, Igor Guardiancich, and David Levi-Faur. 2020. "Modes of Regulatory Governance: A Political Economy Perspective." *Governance* 33 (1): 5–19. https://doi.org/10.1111/gove.12479.

Hall, Peter A., and Rosemary C. R. Taylor. 1996. "Political Science and the Three New Institutionalisms." *Political Studies* 44 (5): 936–57. https://doi.org/10.1111/j.1467-9248.1996.tb00343.x.

Hanretty, Chris, and Christel Koop. 2013. "Shall the Law Set Them Free? The Formal and Actual Independence of Regulatory Agencies." *Regulation & Governance* 7 (2): 195–214. https://doi.org/10.1111/j.1748-5991.2012.01156.x.

Hasegawa, Kōichi. 2004. *Constructing Civil Society in Japan: Voices of Environmental Movements.* Stratification and Inequality Series 3. Melbourne, Victoria: Trans Pacific Press.

Hasegawa, Koichi. 2012. "Facing Nuclear Risks: Lessons from the Fukushima Nuclear Disaster." *International Journal of Japanese Sociology* 21 (1): 84–91. https://doi.org/10.1111/j.1475-6781.2012.01164.x.

Hasegawa, Koichi. 2014. "The Fukushima Nuclear Accident and Japan's Civil Society: Context, Reactions, and Policy Impacts." *International Sociology* 29 (4): 283–301. https://doi.org/10.1177/0268580914536413.

Hogan, John. 2019. "The Critical Juncture Concept's Evolving Capacity to Explain Policy Change." *European Policy Analysis* 5 (2): 170–89. https://doi.org/10.1002/epa2.1057.

Hogan, John, and David Doyle. 2007. "The Importance of Ideas: An A Priori Critical Juncture Framework." *CJP* 40 (4). https://doi.org/10.1017/S0008423907071144.

Homma, Takashi, and Keigo Akimoto. 2013. "Analysis of Japan's Energy and Environment Strategy After the Fukushima Nuclear Plant Accident." *Energy Policy* 62 (November): 1216–25. https://doi.org/10.1016/j.enpol.2013.07.137.

Honda, Hiroshi. 2005. Datsu-Genshiryoku No Undō to Seiji: Nihon No Enerugii Seisaku No Tenkan Wa Kanō Ka [The antinuclear movement and politics: Towards a transformation of Japan's energy policy?]. Sapporo: Hokkaido University Tosho Kankokai.

Honma, Ryu. 2012. Dentsū to Genpatsu Hōdō [Dentsū and nuclear energy coverage]. Tokyo: Akishobo.

Hosono, Gōshi. 2013. Mirai E No Sekinin [Responsibility for the future]. Kadokawa wan tēma nijūichi A-169. Tōkyō: Kadokawashoten.

Hosono, Gōshi, and Shuntarō Torigoe. 2012. Shōgen "Genpatsu Kiki 500-Nichi" No Shinjitsu Ni Torigoe Shuntarō Ga Semaru [Testimony: Torigoe Shuntarō pursues the truth about the "500 days of nuclear crisis"]. Tokyo: Kodansha.

Hughes, Llewelyn. 2015. "Abenomics and Japan's Energy Conundrum." In Pekkanen, Reed, and Scheiner, *Japan Decides*, 199–210. London: Palgrave Macmillan.

Hymans, Jacques E. C. 2011. "Veto Players, Nuclear Energy, and Nonproliferation: Domestic Institutional Barriers to a Japanese Bomb." *International Security* 36 (2): 154–89. https://www.belfercenter.org/publication/veto-players-nuclear-energy-and-nonproliferation-domestic-institutional-barriers.

Hymans, Jacques E. C. 2015. "After Fukushima: Veto Players and Japanese Nuclear Policy." In *Japan: The Precarious Future*, edited by Frank Baldwin and Anne Allison, 110–38. New York: NYU Press.

IAEA. 2003. *Independence in Regulatory Decision Making INSAG-17: A Report by the International Nuclear Safety Advisory Group.* Vienna: IAEA.

IAEA. 2006. *Strengthening the Global Nuclear Safety Regime INSAG-21: A Report by the International Nuclear Safety Group.* Vienna: IAEA.

IAEA. 2007. *Integrated Regulatory Review Service (IRRS) to Japan: IAEA-NSNI-IRRS-2007/01.* Vienna: IAEA.

IAEA. 2011a. "Fukushima Nuclear Accident Update Log: Updates of 12 April 2011." Accessed October 6, 2022. https://www.iaea.org/newscenter/news/fukushima -nuclear-accident-update-log-15.

IAEA. 2011b. "Fukushima Nuclear Accident Update Log: Updates of 12 March 2011." Accessed October 6, 2022. https://www.iaea.org/newscenter/news/fukushima -nuclear-accident-update-log-14.

IAEA. 2011c. *The Great East Japan Earthquake Expert Mission: IAEA International Fact Finding Expert Mission of the Fukushima Dai-Ichi NPP Accident Following the Great East Japan Earthquake and Tsunami.* Accessed October 6, 2022. https:// www-pub.iaea.org/MTCD/Meetings/PDFplus/2011/cn200/documentation /cn200_Final-Fukushima-Mission_Report.pdf.

IAEA. 2012. "IAEA Action Plan on Nuclear Safety." Accessed October 6, 2022. https:// www.iaea.org/sites/default/files/actionplanns.pdf.

IAEA. 2016a. "IAEA Safety Standards: For Protecting People and the Environment." Accessed October 6, 2022. https://www-ns.iaea.org/standards/documents/general.asp.

IAEA. 2016b. *Report of the Integrated Regulatory Review Service (IRRS) Mission to Japan.* Accessed October 6, 2022. https://www.iaea.org/sites/default/files/documents/rev iew-missions/irrs_mission_report_japan_2016.pdf.

IAEA. 2016c. "Press Release: IAEA Mission Says Japan's Regulatory Body Made Fast Progress, Sees Challenges Ahead." Accessed October 6, 2022. http://www.nsr.go .jp/english/cooperation/organizations/IAEA_20160111.html.

IAEA. 2020a. "IAEA Mission Sees Strengthened Safety Inspections in Japan, Encourages Further Enhancement of Regulatory Oversight." Accessed October 6, 2022. https://www.iaea.org/newscenter/pressreleases/iaea-mission-sees-strengthened -safety-inspections-in-japan-encourages-further-enhancement-of-regulatory -oversight.

IAEA. 2020b. *Integrated Regulatory Review Service (IRRS): Follow-up Mission to Japan.* Accessed October 6, 2022. https://www.iaea.org/sites/default/files/documents /review-missions/irrs_japan_follow_up_report.pdf.

Incerti, Trevor, and Phillip Y. Lipscy. 2018. "The Politics of Energy and Climate Change in Japan Under Abe: Abenergynomics." *Asian Survey* 58 (4): 607–34. https://doi .org/10.1525/as.2018.58.4.607.

Infratest. 2011. "ARD Deutschlandtrend April 2011: Eine Umfrage Zur Politischen Stimmung Im Auftrag Der ARD-Tagesthemen Und Drei Tageszeitungen." Accessed October 6, 2022. https://www.infratest-dimap.de/fileadmin/_migrated /content_uploads/dt1104_bericht.pdf.

Ishiwatari, Akira. 2016. "Japan's Nuclear Regulation Against Natural Hazards After the Fukushima Daiichi Accident." https://www.nsr.go.jp/data/000172225.pdf.

Izadi-Najafabadi, Ali. 2015. "Japan's Likely 2030 Energy Mix: More Gas and Solar." Accessed October 6, 2022. https://about.bnef.com/blog/japans-likely-2030-energy -mix-gas-solar/.

JAIF. 2022. "Current Status of Nuclear Power Plants in Japan: As of January 7, 2022." Accessed October 01, 2022. https://www.jaif.or.jp/cms_admin/wp-content/uploads /2022/01/jp-npps-operation20220107_en.pdf.

Japan Times. 2012. "Nuclear Watchdog Autonomy." May 14, online edition. Accessed October 6, 2022. http://www.japantimes.co.jp/opinion/2012/05/14/commentary /world-commentary/nuclear-watchdog-autonomy/#.V3Z5E2df270.

Japan Times. 2014a. "Abe Picks for NRA 'Undermine' Nuclear Watchdog's Independence." June 11, online edition. Accessed October 6, 2022. https://www.japantimes.co.jp /news/2014/06/11/national/abe-picks-nra-undermine-nuclear-watchdogs -independence/.

Japan Times. 2014b. "Volcano Near Sendai Nuclear Plant Is Shaking and May Erupt: Japan Weather Agency." October 24, online edition. Accessed October 6, 2022. http://www.japantimes.co.jp/news/2014/10/24/national/volcano-near-sendai-nuclear-plant-shaking-may-erupt-japan-weather-agency/#.WAUHy3r-m7M.

Japan Times. 2015a. "Court Rejects Injunction Against Sendai Reactors." April 23, morning edition.

Japan Times. 2015b. "Fukui Court Halts Reactor Restarts at Takahama." April 15, morning edition.

Japan Times. 2017. "Japan Downplayed Chernobyl Concerns at G-7 for Energy Policy's Sake: Declassified Documents." December 20, online edition. Accessed October 6, 2022. https://www.japantimes.co.jp/news/2017/12/20/national/history/japan-downplayed-chernobyl-concerns-g-7-energy-policys-sake-declassified-documents/.

Japan Times. 2020. "Shikoku Electric Won't Appeal Injunction over Ikata Nuclear Plant, for Now." Accessed October 6, 2022. https://www.japantimes.co.jp/news/2020/01/27/national/crime-legal/shikoku-electric-wont-appeal-injunction-ikata-nuclear-plant-now/.

Japan Today. 2016. "Abe Says Japan Cannot Do Without Nuclear Power." March 11.

Japanese Energy Industry Expert. 2015. Interview by F. Koppenborg. May 9. Tokyo.

Japanese Nuclear Industry Expert. 2016. Interview by F. Koppenborg. March 31. Tokyo.

Johnson, Chalmers. 1978. *Japan's Public Policy Companies* 60. Washington, DC: American Enterprise Institute for Public Policy Research.

Johnson, Chalmers. 1982. *MITI and the Japanese Miracle: The Growth of Industrial Policy, 1925–1975*. Stanford, CA: Stanford University Press.

Johnson, Chalmers. 1995. *Japan, Who Governs? The Rise of the Developmental State*. New York: Norton.

Jordana, Jacint, Xavier Fernández-i-Marín, and Andrea C. Bianculli. 2018. "Agency Proliferation and the Globalization of the Regulatory State: Introducing a Data Set on the Institutional Features of Regulatory Agencies." *Regulation & Governance* 12 (4): 524–40. https://doi.org/10.1111/rego.12189.

Jordana, Jacint, David Levi-Faur, and Xavier Fernández Fernández-i-Marín. 2011. "The Global Diffusion of Regulatory Agencies: Channels of Transfer and Stages of Diffusion." *Comparative Political Studies* 44 (10): 1343–69. https://doi.org/10.1177/0010414011407466.

Jordana, Jacint, David Levi-Faur, and Xavier Fernández i Marín. 2011. "The Global Diffusion of Regulatory Agencies." *Comparative Political Studies* 44 (10): 1343–69. https://doi.org/10.1177/0010414011407466.

Kahneman, Daniel. 2012. *Thinking, Fast and Slow*. London: Penguin Books.

Kameyama, Yasuko. 2019. *Climate Change Policy in Japan: From the 1980s to 2015*. Routledge Studies in Asia and the Environment 5. London: Routledge.

Kan, Naoto. 2012. Tōden-Fukushima-Daiichi-Genpatsu Jiko: Sōridaijin Toshite Kangaeta Koto [My experience as prime minister during the Fukushima Daiichi nuclear accident]. Gentōsha Shinsho 283. Tokyo: Gentosha.

Keidanren. 2019. "Rebuilding Japan's Electricity System: Electricity Policy to Realize Society 5.0." April 16. Accessed October 6, 2022. http://www.keidanren.or.jp/en/policy/2019/031_outline.pdf.

Kikkawa, Takeo. 2012. Denryoku Kaikaku: Enerugii Seisaku No Rekishi-Teki Daitenkai [Electricity sector reform: Great shifts in the history of energy policy] 2145. Tokyo: Kodansha.

Kingdon, John W. 1984. *Agendas, Alternatives and Public Policies*. Boston: Little, Brown and Company.

Kingston, Jeff, ed. 2014a. *Critical Issues in Contemporary Japan*. London: Routledge.

Kingston, Jeff. 2014b. "Japan's Nuclear Village: Power and Resilience." In Kingston 2014a, 107–19.

Kingston, Jeff. 2016. "Downsizing Fukushima and Japan's Nuclear Relaunch." In *Disasters and Social Crisis in Contemporary Japan: Political, Religious, and Sociocultural Responses*, edited by Mark Mullins and Koichi Nakano, 59–80. Basingstoke, UK: Palgrave Macmillan UK.

Kishimoto, Atsuo. 2017. "Public Attitudes and Institutional Change in Japan Following Nuclear Accidents." In *Policy Shock: Recalibrating Risk and Regulation After Oil Spills, Nuclear Accidents, and Financial Crises*, edited by Edward J. Balleisen, Lori S. Bennear, Kimberly D. Krawiec, and Jonathan B. Wiener, 305–347. Cambridge: Cambridge University Press.

Koppenborg, Florentine. 2016. "Japan's Nuclear Power Plans Don't Add Up." *East Asia Forum*. Accessed January 10, 2020. http://www.eastasiaforum.org/2016/03/03/japans-nuclear-power-plans-dont-add-up/.

Koppenborg, Florentine. 2021. "Nuclear Restart Politics: How the 'Nuclear Village' Lost Policy Implementation Power." *Social Science Japan Journal* 24 (1): 115–35. https://doi.org/10.1093/ssjj/jyaa046.

Krauss, Ellis S., and Benjamin Nyblade. 2005. "'Presidentialization' in Japan? The Prime Minister, Media and Elections in Japan." *British Journal of Political Science*. 35 (2): 357–68. https://doi.org/10.1017/S0007123405000190.

Kurokawa, Kiyoshi, and Andrea Ryoko Ninomiya. 2018. "Examining Regulatory Capture: Looking Back at the Fukushima Nuclear Power Plant Disaster, Seven Years Later." *University of Pennsylvania Asian Law Review* 13 (2): 47–71.

Kushida, Kenji E. 2016. "Japan's Fukushima Nuclear Disaster: An Overview." In Blandford and Sagan 2016, 10–26.

Kushida, Kenji E., and Phillip Y. Lipscy, eds. 2013a. *Japan Under the DPJ: The Politics of Transition and Governance*. Baltimore: The Brookings Institution.

Kushida, Kenji E., and Phillip Y. Lipscy. 2013b. "The Rise and Fall of the Democratic Party of Japan." In *Japan Under the DPJ: The Politics of Transition and Governance*, edited by Kenji E. Kushida and Phillip Y. Lipscy, 3–42. Baltimore: The Brookings Institution.

Kyodo News. 2020. "Costs for Managing Japan's Nuclear Plants to Total 13 Trillion Yen: January 15, 2020." Accessed October 6, 2022. https://english.kyodonews.net/news/2020/01/8722fafaff9b-costs-for-managing-japans-nuclear-plants-to-total-13-trillion-yen.html.

LDP. 2012. "Atarashī Genshiryoku-Kisei-Soshiki Ni Kan Suru Kihonteki Kangae-Kata" [Basic thoughts on the new Nuclear Regulation Agency]. Accessed October 6, 2022. https://www.y-shiozaki.or.jp/contribution/pdf/20120417111607_GUv9.pdf.

LDP Bulletin. 2012. "Shiozaki Yasuhisa Tō Genshiryoku Kisei Soshiki Ni Kan Suru PT Zachō, Chokugeki Intabyū: IAEA No Anzen Kijun No Junshu O Jūshi" [Up-close interview with Shiozaki Yasuhisa, Chairman of the LDP's project team on the Nuclear Regulation Agency: Adherence to IAEA Safety Regulations as a Top Priority]. Accessed October 6, 2022. https://www.y-shiozaki.or.jp/contribution/pdf/20120425165127_sLTp.pdf.

Lévêque, François. 2015. *The Economics and Uncertainties of Nuclear Power*. Cambridge: CambridgeUniversity Press.

Levi-Faur, David, ed. 2011. *Handbook on the Politics of Regulation*. Cheltenham, UK: Edward Elgar Publishing.

Levi-Faur, David. 2011. "The Regulatory State and Regulatory Capitalism: An Institutional Perspective." In Levi-Faur 2011, 662–72.

Lukes, Steven. 1974. *Power: A Radical View. Studies in Sociology*. London: Macmillan.

Lukner, Kerstin, and Alexandra Sakaki. 2011. "Blockade Durch Misstrauen: Das Fukushima-Krisenmanagement." *Zeitschrift für internationale Politik* 19 (80): 49–58.

Lukner, Kerstin, and Alexandra Sakaki. 2013. "Lessons from Fukushima: An Assessment of the Investigations of the Nuclear Disaster." *The Asia-Pacific Journal Japan Focus* 11 (19): 1–29. https://apjjf.org/2013/11/19/Kerstin-Lukner/3937/article.html.

Maeda, Yukio. 2015. "The Abe Cabinet and Public Opinion: How Abe Won Re-Election by Narrowing Public Debate." In Pekkanen, Reed, and Scheiner 2015, 89–102.

Maggetti, Martino, and Yannis Papadopoulos. 2018. "The Principal-Agent Framework and Independent Regulatory Agencies." *Political Studies Review* 16 (3): 172–83. https://doi.org/10.1177/1478929916664359.

Maggetti, Martino, and Koen Verhoest. 2014. "Unexplored Aspects of Bureaucratic Autonomy: A State of the Field and Ways Forward." *International Review of Administrative Sciences* 80 (2): 239–56. https://doi.org/10.1177/0020852314524680.

Mahoney, James, and Kathleen Thelen. 2010. "A Gradual Theory of Institutional Change." In *Explaining Institutional Change: Ambiguity, Agency, and Power*, edited by James Mahoney and Kathleen Thelen, 1–37. Cambridge: Cambridge University Press.

March, James G., and Johan P. Olsen. 1983. "The New Institutionalism: Organizational Factors in Political Life." *American Political Science Review* 78 (3): 734–49. https://doi.org/10.2307/1961840.

McNeill, David. 2014. "Japan's Contemporary Media." In Kingston 2014, 64–75.

Meserve, Richard A. 2009. "The Global Nuclear Safety Regime." *Daedalus* 138 (4): 100–111. https://doi.org/10.1162/daed.2009.138.4.100.

METI. 2006a. "Genshiryoku Rikkoku Keikaku." [Plan for a nuclear state].

METI. 2006b. "Press Release: New National Energy Strategy."

METI. 2010. "The Strategic Energy Plan of Japan: Meeting Global Challenges and Securing Energy Futures—Summary."

METI. 2014. "Strategic Energy Plan." Accessed October 6, 2022. http://www.enecho.meti.go.jp/en/category/others/basic_plan/pdf/4th_strategic_energy_plan.pdf.

METI. 2015. "Long-Term Energy Supply and Demand Outlook." Accessed October 6, 2022. https://policy.asiapacificenergy.org/sites/default/files/Long-term%20Energy%20Supply%20and%20Demand%20Outlook.pdf.

METI. 2018. "Strategic Energy Plan." Accessed October 6, 2022. https://www.enecho.meti.go.jp/en/category/others/basic_plan/5th/pdf/strategic_energy_plan.pdf.

METI. 2021. "The 6th Strategic Energy Plan (Outline)." Accessed October 6, 2022. https://www.enecho.meti.go.jp/en/category/others/basic_plan/pdf/6th_outline.pdf.

Mitnick, Barry M. 2011. "Capturing 'Capture': Definitions and Mechanisms." In Levi-Faur 2011, 34–49.

MOE. 2012. "Genshiryoku-Kisei-I'inkai Junbi Kaigō: Gijiroku" [Preparation meeting for the Nuclear Regulation Authority: Protocol]. Accessed October 6, 2022. http://www.cas.go.jp/jp/genpatsujiko/info/120914gijiroku.pdf.

Mogaki, Masahiro. 2019. *Understanding Governance in Contemporary Japan: Transformation and the Regulatory State*. Manchester: Manchester University Press.

Morgenstern, Christian. 1964. *Palmström-Liedern: The Gallows Songs, Christian Morgenstern's Galgenlieder, a Selection Translated, with an Introduction, by Max Knight.* University of California Press.

Mosley, Layna. 2013. "Introduction. "Just Talk to People"? Interviews in Contemporary Political." In *Interview Research in Political Science*, edited by Layna Mosley, 1–28. Ithaca, NY: Cornell University Press.

NAIIC. 2012. *The Official Report of the Fukushima Nuclear Accident Independent Investigation Commission (NAIIC). Executive Summary.* Accessed October 6, 2022. https://www.nirs.org/wp-content/uploads/fukushima/naiic_report.pdf.

Naito, Kaoru. 2016. "Security Implications of the Fukushima Accident." In Blandford and Sagan 2016, 58–79.

New York Times. 2011. "Japanese Officials Ignored or Concealed Dangers." May 16, 2011, online edition. Accessed October 6, 2022. https://www.nytimes.com/2011/05/17/world/asia/17japan.html.

New York Times. 2012. "Japan Ignored Nuclear Risks, Official Says." February 15, online edition. Accessed October 6, 2022. https://www.nytimes.com/2012/02/16/world/asia/japanese-official-says-nations-atomic-rules-are-flawed.html.

Nihon Keizai Shimbun. 2011. "Rainen Ichi-Gatsu Hōan Teishutsu: Ni-Dankai De Soshiki Saihen" [Bills to be proposed in January next year: Institutional Reorganization to be conducted in two phases]. August 6, morning edition.

North, Douglass C. 1990. *Institutions, Institutional Change and Economic Performance. The Political Economy of Institutions and Decisions.* Cambridge: Cambridge University Press.

NRA. 2013a. *Convention on Nuclear Safety National Report of Japan for 6th Review Meeting August 2013.* Accessed October 6, 2022. https://www.nsr.go.jp/data/000067034.pdf.

NRA. 2013b. *FY 2012 Annual Report.* Accessed October 6, 2022. https://www.nsr.go.jp/data/000067053.pdf.

NRA. 2013c. "Genshiryoku-Kisei-I'inkai Kisha Kaiken-Roku: 09.01.2013" [NRA Board press conference protocol].

NRA. 2013d. "Enforcement of the New Regulatory Requirements for Commercial Nuclear Power Reactors: July 8, 2013." Accessed October 6, 2022. https://www.nsr.go.jp/data/000067212.pdf.

NRA. 2014a. *FY 2013 Annual Report.* Accessed October 6, 2022. https://www.nsr.go.jp/data/000067054.pdf.

NRA. 2014b. "Genshiryoku-Kisei-I'inkai Kisha Kaiken-Roku: Heisei 26 Nen 7 Gatsu 16 Nichi" [NRA Board press conference protocol: July 16, 2014].

NRA. 2014c. "Nuclear Regulation Authority, Japan: Nuclear Regulation for People and Environment."

NRA. 2015a. "Genshiryoku-Kisei-Chō 2015 Nen Saiyō Annai" [NRA Secretariat employment notice 2015].

NRA. 2015b. "Genshiryoku-Kisei-I'inkai-Secchihō Fusoku Dai 6 Jō Dai 2 Kō No Unyō Hōshin Ni Tsuite" [About the "No-return Rule" in Supplementary Article 6 Paragraph 2 of the NRA Establishment Law].

NRA. 2016a. "Genshiryoku-Kisei-Chō 2016 Nen Saiyō Annai" [NRA Secretariat employment notice 2016].

NRA. 2016b. "Outline of Nuclear Regulation of Japan: Reference Documents for the IAEA IRRS Mission." Accessed October 6, 2022. https://www.nsr.go.jp/data/000148578.pdf.

NRA. 2018. *FY 2016 Annual Report.* Accessed October 6, 2022. https://www.nsr.go.jp/data/000253873.pdf.

NRA. 2019a. *FY 2017 Annual Report.* Accessed October 6, 2022. https://www.nsr.go.jp/data/.

NRA. 2019b. "Genshiryoku-Kisei-I'inkai Kisha Kaiken-Roku" [NRA Board press conference protocol]. Accessed October 6, 2022. https://www.nsr.go.jp/nra/kaiken/2019_kaiken.html.

NRA. 2022. *Profile of Commissioners*. Accessed October 6, 2022. https://www.nra.go.jp/english/e_nra/outline/02.html#yamanaka.

NRA Commission. 2012a. "Genshiryoku Kisei Iinkai Kettei: Genshiryoku-Kisei-I'inkaichō Oyobi I'in No Rinri Tō Ni Kakawaru Kōdō Kihan" [NRA Board resolution: Code of conduct for NRA Commission members]. National Diet Library. September 19.

NRA Commission. 2012b. "Genshiryoku Kisei Iinkai Kettei: Gyōsei Kikan No Hoyū Suru Jōhō No Kōkai Ni Kan Suru Hōritsu Ni Motozuku Shobun Ni Kakawaru Shinsa Kijun Tō" [NRA Board resolution: Guidelines for administration based on written documents]. National Diet Library. September 19.

NRA Commission. 2012c. "Genshiryoku-Kisei-I'inkai Kettei: Tōmeisei Chūritsusei Wo Kakuho Suru Tame No Yōken Tō Ni Tsuite" [NRA Board resolution: Guidelines for ensuring transparency and neutrality when consulting external experts]. National Diet Library.

NRA Commission. 2013a. "Shiryō 1: Genshiryoku-Kisei-I'inkai No Soshiki Rinen" [Reference materials 1: NRA mission statement and principles]. National Diet Library. January 9.

NRA Commission. 2013b. "Shiryō 1–2." [Reference material 1–2 (amendment to transparency guidelines)]. National Diet Library. February 6.

NRA Commission. 2014. "Tsuruga-Hatsudensho Shikichinai Hasaitai No Chōsa Nikansuru Yūshikisha Kaigō: 10.12.2014" [Expert meeting for the investigation of fault lines underneath the site of the Tsuruga nuclear power plant]. National Diet Library.

NRA Commission. 2015a. "Heisei 27 Nendo Genshiryoku Kisei Iinkai Dai 30 Kai Kaigi Gijiroku." [Meeting minutes of the NRA Commission's 30th meeting in FY 2015].

NRA Commission. 2015b. "Heisei 27 Nendo Genshiryoku-Kisei-I'inkai No Seisaku Taikei" [Fiscal year 2015 NRA policy goals]. March 11. Accessed October 6, 2022. https://www.nra.go.jp/data/000114540.pdf

NRA Commission. 2015c. "Higashidōri-Genshiryokuhatsudensho Shikichinai Hosaitai No Chōsa Nikansuru Yūshikisha Kaigō" [Expert meeting for the investigation of fault lines underneath the site of the Higashidōri nuclear power plant].

NRA Commission. 2015d. "Kankokubun" [Recommendation text]. Accessed October 6, 2022. https://www.nra.go.jp/data/000150913.pdf

NRA Commission. 2016. "Shika-Genshiryoku Hatsudensho Shikichinai Hosaitai No Chōsa Nikansuru Yūshikisha Kaigō" [Expert meeting for the investigation of fault lines underneath the site of the Shika nuclear power plant].

NRA Commission. 2019. "Heisei 28 Nendo Genshiryoku Kisei Iinkai No Seisaku Taikei" [Fiscal year 2016 NRA policies]. Accessed October 6, 2022. https://www.nra.go.jp/data/000149029.pdf.

NRA HRDC. 2015. "Genshiryoku-Kisei-I'inkai Shokuin No Jinzai Ikusei Ni Kan Suru Torikumi Jōkyō Ni Tsuite" [Current approaches to human resource development for Nuclear Regulatory Authority personnel]. April 28.

NRA Official. 2015. Interview by F. Koppenborg. August 18. Nuclear Regulation Authority, Tokyo.

NRA Official. 2016. Interview by F. Koppenborg. March 14. Nuclear Regulation Authority, Tokyo, Japan.

NRA Official. 2018. Interview by F. Koppenborg. August 14. Nuclear Regulation Authority, Tokyo, Japan.

NRAJapan. 2017. "Fuketa Toyoshi I'inchou Oyobi Yamanaka Shinsuke I'in Shūnin Kaiken" [Entry upon office press conference by Chairman Fuketa Toyoshi and Commissioner Yamanaka Shinsuke]. Accessed October 6, 2022. https://www.youtube.com/watch?v=Pr3BTchcIwI.

Nuclear Politics Expert. 2015. Interview by F. Koppenborg. September 10. Tokyo.

Nuclear Safety Governance Expert. 2015. Interview by F. Koppenborg. September 11. Tokyo.

Pekkanen, Robert. 2003. "Molding Japanese Civil Society: State-Structured Incentives and the Patterning of Civil Society." In *The State of Civil Society in Japan*, edited by Frank J. Schwartz and Susan J. Pharr, 116–34. Cambridge: Cambridge University Press.

Pekkanen, Robert. 2006. *Japan's Dual Civil Society: Members Without Advocates*. East-West Center Series on Contemporary Issues in Asia and the Pacific. Stanford, CA: Stanford University Press.

Pekkanen, Robert J., Steven R. Reed, and Ethan Scheiner, eds. 2015. *Japan Decides 2014: The Japanese General Election*. Basingstoke, UK: Palgrave Macmillan.

Pempel, T.J. 1982. *Policy and Politics in Japan: Creative Conservatism. Policy and Politics in Industrial States*. Philadelphia: Temple University Press.

Pierson, Paul. 2000a. "Increasing Returns, Path Dependency, and the Study of Politics." *The American Political Science Review* 94 (2): 251–67. https://doi.org/10.2307/2586011.

Pierson, Paul. 2000b. "The Limits of Design: Explaining Institutional Origins and Change." *Governance* 13 (4): 475–99. https://doi.org/10.1111/0952-1895.00142.

Poguntke, Thomas, and Paul Webb. 2005. *The Presidentialization of Politics: A Comparative Study of Modern Democracies*. Oxford: Oxford University Press.

Preparation Office Member. 2020. Interview by F. Koppenborg. March 11. Kyoto.

Priest, George L. 1993. "The Origins of Utility Regulation and the "Theories of Regulation" Debate." *Journal of Law and Economics* 36 (1): 289–323. https://www.jstor.org/stable/725477.

Prime Minister's Office. 2011. "Press Conference by Prime Minister Naoto Kan." Accessed October 6, 2022. http://japan.kantei.go.jp/kan/statement/201107/13kaiken_e.html.

Ramseyer, J. Mark. 2019. "The Japanese Judiciary." The Harvard John M. Olin Discussion Paper Series No. 1007. https://papers.ssrn.com/sol3/papers.cfm?abstract_id=3402108.

Ramseyer, J. Mark, and Frances McCall Rosenbluth. 1993. *Japan's Political Marketplace*. Cambridge, MA: Harvard University Press.

Renneberg, Wolfgang. 2011. "The European 'Stress Test' for Nuclear Power Plants: Expertise on Behalf of the Parliamentary Group of the Greens/EFA in the European Parliament." Accessed October 6, 2022. https://www.greens-efa.eu/files/assets/docs/2011_10_22_european_stresstest_final_logo.pdf.

Reuters. 2022. "Japan signals return to nuclear power to stabilise energy supply." August 24, online. Accessed October 6, 2022. https://www.reuters.com/world/asia-pacific/japan-pm-call-development-construction-new-generation-nuclear-power-plants-2022-08-24/.

Sabatier, Paul A., and Christopher M. Weible. 2007. "The Advocacy Coalition Framework: Innovations and Clarifications." In *Theories of the Policy Process*, edited by Paul A. Sabatier. 2nd ed., 189–220. Boulder, CO: Westview Press.

Sagan, Scott D. 2016. "Introduction: Learning from a Man-Made Disaster." In Blandford and Sagan 2016, 3–9.

Sampson, Gary P. 2002. "The Environmentalist Paradox: The World Trade Organization's Challenges." *Harvard International Review* 23 (4). Accessed October 6, 2022. https://go.gale.com/ps/i.do?id=GALE%7CA81390347&sid=googleScholar&v=2.1&it=r&linkaccess=abs&issn=07391854&p=AONE&sw=w&userGroupName=anon%7Ef99dedfd.

Samuels, Richard J. 1987. *The Business of the Japanese State: Energy Markets in Comparative and Historical Perspective*. Ithaca, NY: Cornell University Press.

Samuels, Richard J. 2013. "Japan's Rhetoric of Crisis: Prospects for Change After 3.11." *The Journal of Japanese Studies* 39 (1): 97–120. https://doi.org/10.1353/jjs.2013.0016.

Scalise, Paul J. 2013. "Policy Images, Issue Frames, and Technical Realities: Contrasting Views of Japan's Energy Policy Development." *Social Science Japan Journal* 16 (2): 301–8. https://doi.org/10.1093/ssjj/jyt011.

Scalise, Paul J. 2014. "Who Controls Whom? Constraints, Challenges and Rival Policy Images in Japan's Post-War Energy Restructuring." In Kingston 2014, 92–106.

Scalise, Paul J. 2015. "In Search of Certainty: How Political Authority and Scientific Authority Interact in Japan's Nuclear Restart Process." In *Policy Legitimacy, Science and Political Authority: Knowledge and Action in Liberal Democracies*, edited by Michael Heazle and John Kane, 141–64. Earthscan Science in Society. New York: Routledge.

Schattschneider, Elmer E. 1961. *The Semisovereign People: A Realist's View of Democracy in America*. New York: Holt, Rinehart and Winston.

Schneider, Mycle, and Antony Froggatt. 2013. *World Nuclear Industry Status Report 2013*. Accessed October 6, 2022. https://www.worldnuclearreport.org/IMG/pdf/20130716msc-worldnuclearreport2013-hr-v4.pdf.

Schreurs, Miranda Alice. 2003. *Environmental Politics in Japan, Germany, and the United States*. Cambridge: Cambridge University Press.

Schwartz, Frank Jacob. (1998) 2010. *Advice and Consent: The Politics of Consultation in Japan*. Cambridge: Cambridge University Press.

Segawa, Shirō. 2011. "Genpatsu Hōdō Ha Daihon'ei Happyō Datta Ka. Chō, Mai, Yomi, Nikkei No Kiji Kara Saguru" [Was nuclear power reporting an act of "official statement presentation"? Analyzing articles from the *Asahi, Mainichi, Yomiuri* and *Nikkei Shimbun*]. *Journalism* (255): 28–39.

Shadrina, Elena. 2012. "Fukushima Fallout: Gauging the Change in Japanese Nuclear Energy Policy." *International Journal of Disaster Risk Science* 3 (2): 69–83. https://doi.org/10.1007/s13753-012-0008-0.

Shibata, Tetsuji, and Hiroaki Tomokiyo. 2014. Fukushima-Genpatsu Jiko to Kokumin Yoron [The Fukushima nuclear accident and public opinion]. Tokyo: Energy Review Center.

Shiozaki, Akihisa. 2017. "Political Leadership: The Failure of the DPJ's 'Five Measures.'" In *The Democratic Party of Japan in Power: Challenges and Failures*, edited by Yōichi Funabashi and Kōichi Nakano, 30–51. Nissan Institute/Routledge Japanese Studies Series. London: Routledge Taylor & Francis Group.

Shiozaki, Yasuhisa. 2011a. "Anzen Jitsugen Ni Fukaketsu Na Dokuritsusei, Senmonsei Kakuho" [Ensuring independence and expertise is indispensable for realizing safety]. August 14. Accessed October 6, 2022. https://www.y-shiozaki.or.jp/oneself/index.php?start=130&id=1043.

Shiozaki, Yasuhisa. 2011b. "Kokkai Genpatsu Jiko Chōsa I'inkai" Rippōfu Kara No Chōsenjō [Warning from the "Nuclear Accident Independent Investigation Committee"]. Tōkyō puresu kurabu shinsho 1. Tōkyō: Tōkyōpuresukurabu.

Shiozaki, Yasuhisa. 2011c. "The US-Japan Alliance After 3/11: Remarks by Yasuhisa Shiozaki." Stimson Center, Washington, DC, May 6, 2011. Accessed April 1, 2022. https://www.y-shiozaki.or.jp/en/speech/index.php?start=0&id=55.

Shiozaki, Yasuhisa. 2012. Gabanansu O Seiji No Te Ni, "Genshiryoku Kisei Iinkai" Sōsei E No Tatakai [Returning governance to the hands of politics, the struggle over the creation of the "NRA"]. Tokyo puresu kurabu shinsho 3: Tokyopuresukurabu; Metaburen.

Shiroyama, Hideaki. 2011. "Genshiryokuanyen No Taisei Minaose" [Reforming the Nuclear Safety Administration]. *Nihon Keizai Shimbun,* May 20.

Shiroyama, Hideaki. 2012. "Genshiryoku Anzen Kisei Seisaku: Sengo Taisei No Shūsei Taihensei to Sono Mekanizumu" [Nuclear safety regulation policy: Mechanisms of post-war reforms and reorganization]. In Seisaku Henyō to Seido Sekkei: Seikai Shōchō Saihen Zengo No Gyōsei [Policy change and system design: The administration before and after the reorganization of the ministries, agencies and political circles], edited by Akira Morita and Toshiyuki Kanai, 263–88. MINERVA jinbun shakai kagaku sōsho 179. Tokyo: Mineruva Shobō.

Shiroyama, Hideaki. 2013. "Jichitai No Yakuwari, Seido Seibi Wo" [Preparing a system and a role for independent governance]. *Nihon Keizai Shimbun,* April 22.

Shiroyama, Hideaki. 2015. "Nuclear Safety Regulation in Japan and Impacts of the Fukushima Daiichi Accident." In Ahn et al. 2015, 283–96.

Slater, David. 2011. "Fukushima Women Against Nuclear Power: Finding a Voice from Tohoku." *The Asia-Pacific Journal* 10 (54). Accessed October 24, 2016. http://apjjf .org/-David-H.-Slater/4694/article.html.

Steinmo, Sven, Kathleen Thelen, and Frank Longstreth, eds. 1992. *Structuring Politics: Historical Institutionalism in Comparative Analysis.* Cambridge: Cambridge University Press.

Stigler, George. 1971. "The Theory of Economic Regulation." *The Bell Journal of Economics and Management Science* 2 (1): 3–21. https://doi.org/10.2307/3003160.

Stone, Deborah A. 1989. "Causal Stories and the Formation of Policy Agendas." *Political Science Quarterly* 104 (2): 281. https://doi.org/10.2307/2151585.

Streeck, Wolfgang, and Kathleen Thelen. 2005. "Introduction: Institutional Change in Advanced Political Economies." In *Beyond Continuity: Institutional Change in Advanced Political Economies,* edited by Wolfgang Streeck and Kathleen Thelen, 1–39. Oxford: Oxford University Press.

Suga, Yoshihide. 2021. "Press Conference April 22, 2021." Accessed March 20, 2021. https://www.kantei.go.jp/jp/99_suga/statement/2021/0422kaiken.html.

Tabuchi, Hiroko. 2011. "Japan Leader to Keep Nuclear Phase Out." Accessed October 6, 2022. http://www.nytimes.com/2011/09/03/world/asia/03japan.html?_r=0.

Tateishi, Masata. 2015. "Genpatsu No Taishinsei Anzen Mondai to Shinkiseikijun" [The issue of earthquake-resistant nuclear power plants in the new safety standards]. In Hyōryūsuru Genpatsuryoku to Saikadō Mondai: Nihon-Kagaku-Shakai Gidai 35 Kai Genshiryoku Hatsuden Mondai Zenkoku Shinpojiumu (Kanazawa) Yori [Nuclear power in drift and the issue of restarts: The 35th Japanese Scientists' Meeting, from the nationwide symposium (Kanazawa) on issues relating to nuclear power generation], edited by Nihonkagakushakaigi-Genshiryokumondaikenkyūiinkai, 95–104. Tokyo: Honnoizumi.

TEPCO. 2019a. "Fukushima Daini Nuclear Power Station to Be Decommissioned: Press Release July 31, 2019." Accessed October 6, 2022. https://www.tepco.co.jp/en /hd/newsroom/press/archives/2019/hd_190731_01-e.html.

TEPCO. 2019b. "Report to Kashiwazaki City Mayor Sakurai on the 'Basic Approach to the Recommencement of Operation and Decommissioning of the Kashiwazaki-Kariwa Nuclear Power Station': Press Release August 26, 2019." Accessed October 6, 2022. https://www.tepco.co.jp/en/hd/newsroom/press/archives/2019/report -to-kashiwazaki-city.html.

TEPCO. 2020. "Higashidori Nuclear Power Station: Construction Plan Overview." Accessed October 6, 2022. https://www.tepco.co.jp/en/hd/ourbusiness/nuclear /higashi-dori/index-e.html.

The Diplomat. 2018. "Ex-TEPCO Executive Downplays Role in Fukushima Nuclear Meltdown: Three TEPCO Leaders on Trial for Allegedly Delaying Tsunami Preparation Measures." October 31. Accessed October 6, 2022. https://thediplomat.com/2018/10/ex-tepco-executive-downplays-role-in-fukushima-nuclear-meltdown/.

Thelen, Kathleen. 1999. "Historical Institutionalism in Comparative Perspective." *Annual Review of Political Science* 2 (1): 369–404. https://doi.org/10.1146/annurev.polisci.2.1.369.

Togni, Giulia de. 2022. *Fall-Out from Fukushima: Nuclear Evacuees Seeking Compensation and Legal Protection After the Triple Meltdown.* Nissan Institute/Routledge Japanese Studies. Abingdon, UK: Routledge.

Tokyo Shimbun. 2015a. "Genpatsu Anzenhi 2.3 Chōenzō 13 Nen-Shinkijungo, Yuragukeizaisei" [Nuclear safety costs increase by 2.3 trillion yen following 2013 safety standards, question economic viability]. May 17, morning edition.

Tokyo Shimbun. 2015b. "Takahama Saikadō Mitomezu" [No approval of Takahama restart]. April 15, morning edition.

Tokyo Shimbun. 2016. "Genpatsu Unten Enchō O Zentei, Keizaishō Dengen Kōsei-Hi 20% Iji" [Prolonged operation of nuclear plants set as a precondition, METI minister (says nuclear power) will continue to constitute 20% of power supply]. January 16, morning edition.

Toyo Keizai. 2018. "Iitatemura kara kangaeru nihon no seiji no kekkan to shohousen: kurokawa kiyoshi-ha ga Tanaka shun'ichi-ha o tazune, hanashiatta" [Kurokawa Kiyoshi visits Tanaka Shun'ichi for a chat about the flaws of Japanese politics and remedies]. June 17.

United Nations. 2002. "Press Conference on Chernobyl by Office for Coordination of Humanitarian Affairs." Accessed October 6, 2022. https://www.un.org/press/en/2002/OCHAPC.doc.htm.

Upham, Frank K. 1987. *Law and Social Change in Postwar Japan.* Cambridge, MA: Harvard University Press.

UPI. 2011. "IAEA Sees Slow Nuclear Growth Post Japan." September 23. Accessed October 6, 2022. https://www.upi.com/Energy-News/2011/09/23/IAEA-sees-slow-nuclear-growth-post-Japan/87041316777856/?ur3=1.

Vande Putte, Jan, Kendra Ulrich, and Shaun Burnie. 2015. "The IAEA Fukushima Daiichi Accient Summary Report: A Preliminary Analysis." Accessed October 6, 2022. https://www.greenpeace.org/static/planet4-international-stateless/2018/01/36e90444-36e90444-iaea-analysis-by-gp-20150528.pdf

Vivoda, Vlado. 2014. *Energy Security in Japan: Challenges After Fukushima. Transforming Environmental Politics and Policy.* Farnham: Taylor & Francis.

Vivoda, Vlado, and Geordan Graetz. 2014. "Nuclear Policy and Regulation in Japan After Fukushima: Navigating the Crisis." *Journal of Contemporary Asia* 45 (3): 1–20. 10.1080/00472336.2014.981283.

Vogel, David. 2012. *The Politics of Precaution: Regulating Health, Safety, and Environmental Risks in Europe and the United States.* Princeton, NJ: Princeton University Press.

Vogel, Ezra F. 1999. *Japan as Number One: Lessons for America.* Lincoln, NE: iUniverse.com, Inc.

Vogel, Steven Kent. 2006. *Japan Remodeled: How Government and Industry Are Reforming Japanese Capitalism.* Ithaca, NY: Cornell University Press.

Watanabe, Rie. 2011. *Climate Policy Changes in Germany and Japan: A Path to Paradigmatic Policy Change.* Routledge Research in Comparative Politics 41. Abingdon, UK York: Routledge.

Watanabe, Rie. 2013. "Nuclear Policy Change in Japan After Fukushima: Beliefs, Interests, and Positions." Presentation at FFU Seminar at University of Salzburg, 2013. Accessed October 6, 2022. https://www.polsoz.fu-berlin.de/polwiss/forschung/systeme/ffu/veranstaltungen/termine/downloads/13_salzburg/Watanabe-Salzburg-2013.pdf

Weiß, Tobias. 2019. Auf der Jagd nach der Sonne: Das journalistische Feld und die Atomkraft in Japan. 1st ed. Studien zur politischen Soziologie 36. Baden-Baden: Nomos.

Weiss, Tobias. 2020. "Journalistic Autonomy and Frame Sponsoring. Explaining Japan's "Nuclear Blind Spot" with Field Theory." *Poetics* 80 (2): 101402. https://doi.org/10.1016/j.poetic.2019.101402.

Wenisch, Antonia, and Oda Becker. 2012. "Critical Review of the EU Stress Test Performed on Nuclear Power Plants: Study Commissioned by Greenpeace."

Wiemann, Anna. 2018. *Networks and Mobilization Processes: The Case of the Japanese Anti-Nuclear Movement After Fukushima.* Monographien aus dem Deutschen Institut für Japanstudien Band 61. Munich: Iudicium.

Wilson, James Q. 1980. "The Politics of Regulation." In *The Politics of Regulation*, edited by James Q. Wilson, 357–94. New York: Basic Books Inc.

Wilson, James Q. 1989. *Bureaucracy: What Government Agencies Do and Why They Do It.* New York: Basic Books Inc.

Worldwatch Institute. 2013. "Nuclear Power's Uncertain Future: New Worldwatch Institute Analysis Examines the State of the Global Nuclear Energy Industry."

Wyss, Gregory D. 2016. "The Accident That Could Never Happen: Deluded by a Design Basis." In Blandford and Sagan 2016, 29–57.

Yomiuri Shimbun. 2011. "Kankyōshō Gaikyoku De Kecchaku E: Seifu Hōshin, Toppu Wa Yūshikisha" [Toward (the establishment of) an external bureau to the Ministry of the Environment: Government plan to put experts at the top]. August 12, morning edition.

Zakowski, Karol. 2015. *Decision-Making Reform in Japan: The DPJ's Failed Attempt at a Politician-Led Government.* Routledge Contemporary Japan Series 58. Abington, UK: Routledge.

Index

Page numbers in **bold** and ***bold italicized*** refer to figures and tables, respectively
NP stands for nuclear power
NPP stands for nuclear power plant

backup systems, NRA requirement, 167
business environment, 149, 159, 177
deregulation, 40
electricity generation by fuel (2000–2016), **45**
investment strategies, 146–49, *147*, 159, 162
and NP support, dissatisfaction, 134–35
opposition to government controls, 35
shaping of, 29–32
systems and responsibilities, 30, 187n2
See also costs, NP
emergency response
Cabinet Office, sanctions, 95
expanded planning, 118–19
flaws in, pre-3.11, 128
guidelines, 90–91, 119
legal framework, 72
measures, 65–66, 113
NRA Commission, 88
NRA responsibilities, 72, 95, 98, 156–57
preparedness, 25, 178
regulations, **116**, 128, 155, 156–57
staffing, 105
support system, 42
TEPCO headquarters, 57
Energy and Environment Council, 61
Energy and Environment Policy Committee
(EEC), public involvement in NP
policy, 73
energy policy, 6, 41, 48, 60, 72, 135, 183
energy security, 43, 48
energy strategy, post-3.11, 134
environment, 6, 8, 43, 58, 128–29, 133, 164,
182
Environmental Dispute Coordination
Commission, 171
European Union (EU), 4, 52, 128
expertise
accident response ministries, lack of, 57
international, 62–64
lack of, 34, 49, 57, 86–87, 103, 108, 109–11,
168–69
political power distribution, role in, 10,
184–85
See also independence; training
external safety agency, establishment, 69–71
See also LDP and Kōmeitō opposition;
Ministry of Environment

Federation of Electric Power Companies
(Denjiren). *See* Denjiren
feedback mechanisms, 14, 16, 177, 179, 180
formal and informal procedures and rules, 13,
28, 132

formal independence. *See* independence
forty-year rule
Abe government pressure to abandon, 157,
167
absence in NRA Establishment Act, 72
extension, 104, 116, 124, 135, 136
LDP and Kōmeitō counterproposal, 70
Noda Cabinet draft bill, 63
NRA Commission, adoption, 77, 91, 115–16
framing analysis, 23, 53, 67
Fugen, advanced thermal reactor accident, 37
Fuketa Toyoshi, 103
end of term, 172
on improving skills and safety culture, 102,
126–27
NRA board member, 75, 76, 88, 120
on NRA nuclear safety commitment, 130
second NRA Commission chair, and
continuity, 92, 109–10, 141, 142, 167
Fukui District Court, 154, 156
Fukushima Daiichi NPP. *See* Daiichi NPP
Fukushima Daini NPP. *See* Daini NPP
Fukushima nuclear accident, 1–2, 52–53,
64–65, 69, 95
Funabashi report (2012), 64
Furukawa Motohisa, 72–73

Genkai NPP, expansion, 37
Germany, risks and benefits of nuclear power
assessments, 2
government intervention system, 34
Greenpeace Japan, nuclear safety issues
posting, 149
Gustavo, Caruso, lessons learned, 62

Hamaoka NPP, 37, 39, 44, 146
Hashimoto reforms, 15, 41–42, 49, 82–83
Hatamura report (2012), 64–65
Hatoyama Yukio, 60
Hayashi Motō, on NP lifetime extension, 136
Higashidōri NPP, seismic fault line investigation, 123
high-risk high-return technologies, and
politics, 10–11, 183–84
Higuchi Hideaki, presiding judge, NRA safety
standards critique, 154
Hiroshima High Court, 154, 155
historical institutionalism, 13, 17, 21, 173
scholars' language of costs, 24, 51, 179
studies, long-term perspective, 84
temporality, focus on, 26, 165
view of institutional change, 12, 13
Hitachi, 8, 36, 44, 161